Gene Expressions in Response to Diseases, Abiotic Stresses and Pest Damage of Horticultural Products

Gene Expressions in Response to Diseases, Abiotic Stresses and Pest Damage of Horticultural Products

Guest Editors

Giuliana Maddalena
Annamaria Mincuzzi
Francesca Garganese
Antonio Ippolito

MDPI

Basel • Beijing • Wuhan • Barcelona • Belgrade • Novi Sad • Cluj • Manchester

Guest Editors
Giuliana Maddalena
Department of Agricultural and Environmental Sciences
University of Milan
Milan
Italy

Annamaria Mincuzzi
Department of Soil, Plants and Food Sciences
University of Bari Aldo Moro
Bari
Italy

Francesca Garganese
Department of Soil, Plants and Food Sciences
University of Bari Aldo Moro
Bari
Italy

Antonio Ippolito
Department of Soil, Plants and Food Sciences
University of Bari Aldo Moro
Bari
Italy

Editorial Office
MDPI AG
Grosspeteranlage 5
4052 Basel, Switzerland

This is a reprint of the Special Issue, published open access by the journal *Horticulturae* (ISSN 2311-7524), freely accessible at: https://www.mdpi.com/journal/horticulturae/special_issues/Gene_Expressions_Response.

For citation purposes, cite each article independently as indicated on the article page online and as indicated below:

Lastname, A.A.; Lastname, B.B. Article Title. *Journal Name* **Year**, *Volume Number*, Page Range.

ISBN 978-3-7258-5169-0 (Hbk)
ISBN 978-3-7258-5170-6 (PDF)
https://doi.org/10.3390/books978-3-7258-5170-6

Contents

About the Editors

Giuliana Maddalena

Giuliana Maddalena received her master's degree in Plant Production and Protection from the University of Milan in 2013. She earned her PhD in Agriculture, Environment and Bioenergy at the same university in 2017, focusing on the phenotypic characterization of the interaction between *Plasmopara viticola* (the causal agent of grapevine downy mildew) and *Vitis vinifera*, including epidemiological studies and genetic variability of the pathogen population. Since January 2025, she has been a tenure-track researcher (RTT) in the scientific-disciplinary sector AGRI-05/B (Plant Pathology) at the Department of Agricultural and Environmental Sciences, Production, Territory, Agroenergy, University of Milan. Previously, she held post-doctoral and research-fellow positions at the same department, where she conducted research on sustainable management of grapevine diseases, particularly downy mildew, using epidemiological models and resistance management strategies. She has contributed to several key projects including "Grape4vine" funded by Fondazione Cariplo, the sustainable management of downy mildew in collaboration with the Franciacorta Consortium, the H2020 "Novel Pesticides for a Sustainable Agriculture (NoPest)", and regional initiatives such as the PSR 2014-2020 in Lombardy. She is a member of the Italian Association for Plant Protection (AIPP) and the Italian Society of Plant Pathology (SIPaV).

Annamaria Mincuzzi

Annamaria Mincuzzi received her master's degree in Environmental Biology and PhD in Plant Pathology from the University of Bari in 2015 and 2019, respectively. She was a post-doctoral researcher at the Department of Soil, Plant, and Food Sciences (DiSSPA) at the above-mentioned Italian university until 2023; her research activity includes the alternative control means (chemical, physical and biological) of postharvest fungal diseases of fruits, particularly of pomegranates. The evaluation of new bioactive polyphenolic compounds extracted by plants, the host–pathogen interactions, and molecular techniques to identify phytopathogenic microorganisms are her chief research topics. In 2024 and 2025, she attended the Bari headquarters of the Institute for Sustainable Plant Protection (IPSP) of the National Research Council (CNR) as senior post-doctoral researcher. She carried out olive micropropagation, testing the effectiveness of beneficial bacteria for growth promotion; furthermore, she worked on Olive Quick Decline Syndrome (OQDS) caused by *Xylella fastidiosa* in Apulian olive trees. In October 2025, she will be a tenure-track researcher at the University of Bologna dealing with Postharvest Plant Pathology. She has collaborated with the Biosciences, Biotechnologies and Environment Department (DBBA) of the University of Bari, focusing on species identification and bioactive extraction from macroalgae. She took part in national and international projects (Oltre.Bio, StopMedWaste, Euphresco, IntegrOliv, BioVexo) and conferences both as a speaker, auditor, and chairperson. She published her research results in indexed international journals, also collaborating as a reviewer and as a member of the topical advisory panel. She is a member of the 'Associazione Italiana per la Protezione delle Piante' (AIPP) and of the 'International Society for Horticultural Science' (ISHS).

Francesca Garganese

Francesca Garganese graduated with a master's degree in Industrial and Environmental Biotechnologies and obtained a PhD in Crop Protection from the University of Bari Aldo Moro, Bari, Italy. She is a research fellow in Entomology at the Department of Soil, Plant, and Food Sciences at the University of Bari, involved in the Agritech PNRR project. She achieved the national scientific qualification of Associate Professor in the Italian higher education system, in the disciplinary field of 07/D1—plant pathology and entomology, on February 9, 2023. During her PhD, she worked on the epidemiology, characterization, and control of citrus-associated *Alternaria* species, as well as the cloning of *Alternaria alternata* genes involved in mycotoxin production and the knockdown of genes using siRNA (RNA interference). Her research interests encompass the study of insect-borne plant pathogens, the insect microbiota, insect–plant interactions, environmental IPM, vector pest damage, insect bionomics, and innovative molecular techniques, particularly MinION mass sequencing (Oxford Nanopore Technologies).

Antonio Ippolito

Antonio Ippolito is a full Professor of Plant Pathology. A member of the PhD course in Crop Protection since 1993, he was the coordinator of the PhD program in Biodiversity, Agriculture and Environment from 2009 to 2021; from 2006 to 2015 and since 2024, he has been the head of the teaching and research activities at the 'P. Martucci' educational–experimental center of the Faculty of Agriculture of Bari. He is the academic coordinator for student mobility of the Erasmus program with the "Namik Kemal" University of Tekirdag, and the "Uludag" University of Bursa, Turkey. He was involved in the design and coordination of the activities of the university master's degree "Post-harvest Management of Fresh Fruit and Vegetable Products" - PON Scientific Research, Technological Development, Higher Education, 2000/2006 - Axis III - Measure III.4 A. His scientific activity is documented by over 400 papers in national and international journals (138 with IF; H-index=45), as well as 26 chapters in international and national volumes and various articles and abstracts in conference proceedings, and he has treated various aspects of plant pathology of Mediterranean crops. For over 30 years, he has been carrying out research on the use of alternative means to chemicals for controlling postharvest rot in fresh fruit and vegetables. He has been and is the head of research units in research projects funded by the MIUR, as well as regional and interregional projects and projects funded by the European Union, and private industries. He is a member of the editorial board of several international journals. In 2015, he was the convener of the "III International Symposium on Postharvest Pathology", Bari, on 7-11 June. From 2015 to 2019, he served as Chair of the International Workgroup "Biological Control of Postharvest Diseases" of the International Society for Horticultural Science. He is a member of several international associations in the field of plant pathology.

horticulturae

MDPI

Editorial

Special Issue: Gene Expressions in Response to Diseases, Abiotic Stresses, and Pest Damage of Horticultural Products

Francesca Garganese [1,*], Giuliana Maddalena [2], Antonio Ippolito [1] and Annamaria Mincuzzi [1,3,*]

[1] Department of Soil Sciences, Plants and Food (DISSPA), University of Bari Aldo Moro, Via Amendola 165/A, 70126 Bari, Italy; antonio.ippolito@uniba.it
[2] Department of Agricultural and Environmental Sciences (DiSAA), University of Milan, Via Celoria 2, 20133 Milano, Italy; giuliana.maddalena@unimi.it
[3] Department of Agricultural and Food Sciences (DISTAL), University of Bologna, 40126 Bologna, Italy
* Correspondence: francesca.garganese@uniba.it (F.G.); annamaria.mincuzzi@uniba.it (A.M.)

1. Introduction

Horticultural crops are continuously exposed to pathogens, pests, and abiotic stress, which collectively result in significant economic losses. Over time, plants have evolved sophisticated defense mechanisms to counteract both pathogen and insect attacks, as well as to face abiotic stresses. The identification of genes involved in defense responses and stress tolerance is fundamental for selecting resistant crops, which could enable the sustainable management of diseases and damages. Gene expression is subject to complex regulation, influenced not only by developmental and tissue-specific factors, but also by both biotic and abiotic stimuli in the environment. The fluctuation in expression levels, or the gene expression profile, provides essential insights into the biological functions of genes. Comprehensive expression profiles established under various experimental conditions could facilitate the understanding of gene functions. Similarities in expression profiles between genes are an effective index for predicting gene function and mechanisms of gene expression regulation [1]. If two genes have similar expression profiles, it is reasonable to expect that they have the same biological function, even in different plant species. In addition, similar expression profiles suggest that two genes are regulated by the same mechanism (e.g., a transcription factor (TF) and a cis-element). Therefore, genes with highly similar expression profiles are likely to have the same biological functions and regulatory mechanisms. Transcriptomics enables the simultaneous analysis of gene expression in plants, pathogens, and pest interactions (Figure 1).

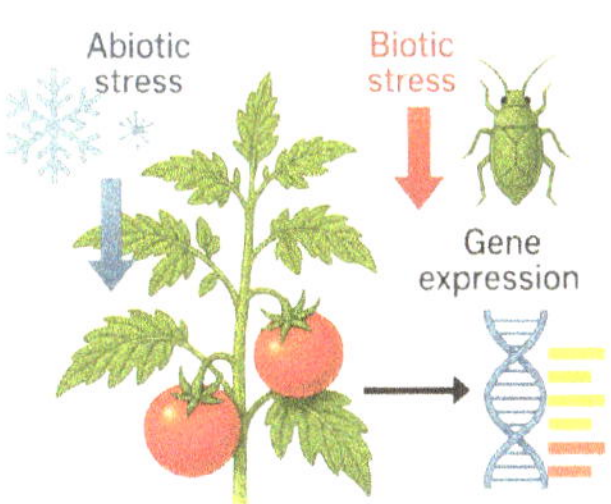

Figure 1. A tomato plant under abiotic (blue arrow) and biotic stress (red arrow) is shown; to understand the mechanisms involved in stress, studying gene expression is relevant and fundamental. Image generated with the help of Chat GPT (https://chatgpt.com (accessed on 26 June 2025).

 https://doi.org/10.3390/horticulturae11080920

2. Overview of Published Articles

Several gene families are involved in plant responses to biotic and abiotic stresses, becoming objects of research aiming to enable breeding improvement. The first two manuscripts address this issue by examining the evolutionary relationship, conserved motif composition, gene structure, chromosomal location, cis-regulatory elements within the promoter region, and primary expression patterns in various plant tissues (stem, root, leaf, tendril, and flower) and in response to both abiotic and biotic stresses.

In Contribution 1, the organ-specific expression patterns and stress-responsive gene expression patterns of 24 eggplant *DIR* family genes were investigated. Directing proteins (*DIR*) play essential roles in lignin and lignan pathway biosynthesis, protective responses, and secondary metabolism, as well as in plant disease resistance. The results identified 24 *DIR* genes in eggplant, divided into three subfamilies (*DIR-a*, *DIR-b/d*, and *DIR-e*). Among the identified cis-elements, some were involved in hormone response, stress defense, and the expression of plant tissues, suggesting a correlation with plant development and plant response to abiotic and biotic stresses. Specifically, the expression of each *SmDIR* gene displayed organ-specificity and typical expression (down- or upregulation) according to cultivar sensitivity. The *SmDIR22* gene was among the candidates for chief association with stress response, being differentially expressed in response to abiotic and biotic stresses. The results of the study will provide a foundation for future research on the biological functions of the *SmDIR22* gene, which is promising for the molecular breeding of resistant eggplant.

In Contribution 2, the authors characterized the SIMILAR TO RCD-ONE (*SRO*) gene family in two-leaf-stage seedlings of Chinese Long cucumber; SRO is a family of small, plant-specific proteins regulating numerous biological processes, including physiological development and stress reactions. These genes, which encode transcription factor (TF) proteins CsSRO1, CsSRO2, CsSRO3, and CsSRO4, were successfully isolated, cloned, and found to be deeply induced under drought and salt stresses, displaying different expressions in sensitive and resistant cucumber plants. Expression profiling revealed that the four *CsSRO* genes were differentially expressed in various tissues and under different stress conditions, indicating their involvement in plant physiological processes and defense mechanisms. The *SRO* family genes were relatively conservative, with the authors revealing conserved motif arrangements within each subgroup and highlighting differences between monocot- and dicot-specific patterns. Except for CsSRO2, which is located in both the nucleus and mitochondria, the other *SRO* genes were found in the nucleus. In the future, the four *SRO* genes could be considered potential candidates for breeding stress-tolerant cucumber varieties. Being related to phytohormone signaling, they could also be involved in defense responses to biotic stresses.

Plants have evolved wide phytohormone signaling pathways to improve survival in response to drought stress, such as the auxin (IAA), cytokinin (CK), abscisic acid (ABA), jasmonic acid (JA), salicylic acid (SA), ethylene (ET), and gibberellin (GA) pathways [2,3]. Genes and pathways involved in stress response were physiologically and transcriptionally studied in the Contribution 3, and *Bletilla striata*, a widely used medicinal plant, was found to be highly susceptible to drought stress during the planting chain. In plant tissues, biotic and abiotic stresses caused ROS accumulation and early defense relied on antioxidants such as catalase (CAT), peroxidase (POX), superoxide dismutase (SOD), glutathione sulfotransferase (GST). Indeed, these pathways were upregulated: The upregulation of *SnRK2* and *SLAH2* in *B. striata* suggests that drought stress promotes the closure of stomata to prevent water loss. Additionally, SnRK2 can activate downstream genes, such as *ABF*; its overexpression in *Arabidopsis* resulted in ABA hypersensitivity and drought tolerance. Also, the upregulation of *SnRK2* can activate different cation or anion channels, such as S-type anion channel 3 (SLAH3), involved in forcing stomata closure [4]. Enhanced O_2 production also protected the photosystem under drought stress. Furthermore, the drought-induced increase in dehydrin (DHN) gene expression proved ROS scavenging to be the activation of mechanism of drought tolerance [5]. These results indicate that *B. striata* leaves reduce drought-induced oxidative damage by activating the expression of SOD, CAT, and POX, which scavenge ROS, confirming this mode of action.

When analyzing relative gene expression, a preliminary and foundational step is the employment of robust and reliable housekeeping genes. The ideal reference gene is continuously expressed in all tissues across various conditions, and this issue was evaluated in the Contribution 4. Considering eight frequently used housekeeping genes (*EF1α*, *ACT*, *SKD1*, *YLS8*, *UBQ*, *GAPDH*, *TUB*, and *WDP*), the authors evaluated the expression of these candidate reference genes in pear (*Pyrus bretschneideri* Rehd.) leaves subjected to hormonal (SA, MeJA, ETH, and ABA), biotic (*Venturia nashicola*), and abiotic stresses (cold, heat, water deficit). No one gene was consistently expressed across all stress conditions; however, *WDP* was relatively stable except for under methyl jasmonate stress, where ACT had the most stable expression. As they are robustly expressed in plant tissues under different stresses, it is suggested that only *EF1α* and *GAPDH* should be used as reference genes in treatments. These results provide valid and updated inputs for selecting housekeeping genes in pears exposed to biotic and abiotic stresses in the future.

In view of sustainable agriculture, the Contribution 5 investigated the antimicrobial effects related to mycorrhizal-induced resistance (MIR) in tropical chili pepper seedlings differing in mycorrhizal response and pathogen resistance. In their work, the impact of mycorrhizae on the relative expression of genes involved in jasmonic acid (JA)- and salicylic acid (SA)-mediated signaling pathways was studied in plants challenged by *Ralstonia solanacearum*. The effect of mycorrhizae on the phenotypic traits of tropical chili peppers was also studied. In ten tropical chili pepper genotypes, mycorrhiza treatment plus infection with *R. solanacearum* (M1R1) increased the relative gene expression of JA, which was higher in genotypes with high mycorrhizal responsiveness than in those infected with *R. solanacearum* (M0R1). A similar result was found for the resistant-control cultivar. Based on phytohormone balance and relative expression, the results showed a significant increase in the relative gene expression of SA and JA pathways; particularly, the latter enhanced induced systemic resistance (ISR) caused by interactions between the plant and beneficial microorganisms. Generally, the relative gene expression of SA was associated with pathogen resistance; however, JA expression was deeply related to the tropical chili peppers' response to both mycorrhizal colonization and pathogen resistance. The rise in the JA and SA signaling pathways was influenced by the resistance properties of the tropical pepper genotypes to *R. solanacearum*, differences which affected plant phenotypes. The effectiveness of symbiosis depended on the mycorrhizal species and host genotype. It is probable the beneficial effects due to mycorrhizae and observed under biotic stressors could be spread to abiotic stress conditions as well.

The genus *Dianthus*, a member of the family Caryophyllaceae [6], is an ornamental flower with production characteristics similar to those of *Chrysanthemum* and *Rosa*. In Contribution 6, gene expression profiles were studied after the exposure of a *Dianthus hybrida* flower to tactile stimuli. Touch stimulation induced concentration enhancement of free cytosolic Ca^{2+} and activation of Ca^{2+}-binding proteins involved in the regulation of downstream cascades through phosphorylation signaling; the concentration of JA also improved. Touch stimuli induced the upregulation of genes encoding protein kinase, xyloglucan endotransglucosylase, calmodulin, and components involved in JA biosynthesis and signaling, as well as the downregulation of photosynthetic genes, and the significant of phytohormones was highlighted. The authors applied RNA-seq methods and analyses to assess the molecular mechanisms involved in the touch stimulus responses of *D. hybrida* flower buds. Among the tested genes, those encoding serine/threonine kinases were only found in flower buds displaying an organ-specific response to stimulation. This knowledge can be applied in future breeding to enhance flower morphology and develop new varieties.

Pepper production is affected by abiotic and biotic stress [7,8]. In particular, the fungus *Verticillium dahliae* poses a serious threat to global pepper production. With this in mind, Contribution 7. physiologically and transcriptomically compared detached pepper leaves at three time-points after inoculation with *V. dahliae* to define resistant plants in the eighth contribution to this Special Issue. Susceptible and resistant cultivars shared increases in the same signaling pathways related to the transduction of plant–pathogen interaction, MAPK, and phytohormone signaling. The resistant cultivar MS72 exhibited higher chlorophyll levels and a lower malondialdehyde content after inoculation compared to the susceptible cultivar MS66. The research indicates that infection with *V. dahliae* causes cell membrane lipid peroxidation in MS66. Moreover, MS72 maintained high chlorophyll levels and a low MDA content, indicating its resistance to *V. dahliae*. Thirty-six hub genes contribute to the resistance response, such as the transcription factor bHLH93, defense-like protein 1, and miraculin-like. Simultaneously, pathways involved in setting up the cell wall, phenylpropanoid biosynthesis, and photosynthesis (chlorophylls reduced over 70% in susceptible cultivars) were differentially expressed and regulated in resistant cultivars, improving pepper resistance. The key genes and regulatory mechanisms may serve as targets for improving pepper resistance in breeding programs and selecting improved varieties.

A different approach to controlling the oomycete *Phytophthora infestans*, responsible for the late blight disease of the Solanaceae family, was used in the seventh contribution by Contribution 8: virus-induced gene silencing (VIGS) was used to deliver homologous sequences. PiLLP, the ortholog of plant loricrin-like protein (LLP), was silenced, arranging the recombinant vector containing both tobacco mosaic virus and LLP. Thus, RNAinterference (RNAi) could be expressed in the A1 mating type, preventing sexual reproduction and reducing losses and management costs related to late blight disease. RNA interference (RNAi) is a powerful tool for studying gene function that can be deployed through constitutive transformation or transient expression, such as in VIGS experiments. The study provides the first evidence of downregulation of a constitutive gene in *P. infestans* using a recombinant vector based on a plant virus.

Good agronomic practices, such as fertilization, could reduce symptoms caused by biotic and abiotic stress. Contribution 9 provide a comprehensive study on iron deficiency chlorosis limiting the productivity of 'Yali' pears in alkaline soils. Studies on photosynthesis, root vitality, enzymatic activity, soil physicochemical properties, iron and citrate contents, and relative expression of genes linked to iron absorption and transport could contribute to controlling homeostasis regulation and lead to selecting iron-efficient cultivars. This study systematically investigated the physiological and molecular responses to various degrees of iron deficiency, focusing on the roles of *PbFRO2* (*Pyrus bretschneideri Ferric Reductase Oxidase* 2), *PbIRT1* (*P. bretschneideri Iron-Regulated Transporter* 1), and *PbCS2* (*P. bretschneideri Citrate Synthase* 2) in iron uptake. In this paper the authors compared physiological and molecular iron deficiency and leaf/root chlorosis of different severity (moderate, normal, and severe) over time in 'Yali' pear. The bivalent iron content was significant and inversely related to severe chlorosis; furthermore, maximum photochemical efficiency was reduced due to iron stress. The results demonstrated that moderate iron deficiency upregulated *PbFRO2*, enhanced root ferric reductase (FCR) activity, and promoted Fe^{3+} reduction and Fe^{2+} transport, while severe deficiency suppressed the expression of these genes. To evaluate these metabolic pathways, the relative expression of corresponding genes was evaluated and *PbFRO2* was induced by chlorosis to improve radical iron uptake. Globally, normal and moderate chlorotic leaves exhibited similar physiological and molecular features, despite the fact that severe chlorosis induces significant changes. The described mechanisms were led by the *PbFRO2* gene, with the differences related to the season and displaying stage-specific responses to chlorosis according to the "gene–metabolite–phenotype" pattern. This hub gene is suitable for studying adaptive responses, for reducing chlorosis, and for defining potential aims for iron biofortification. Above all, this study proposes *PbFRO2* as a potential molecular target for breeding iron-efficient pear cultivars.

3. Conclusions

Most of the papers in this Special Issue share a common scientific approach, highlighting the morphological, molecular, and metabolic processes that different plant species exhibit in response to biotic and abiotic stresses. These stresses cause a reduction in chlorophyll amounts, chiefly appreciable in leaf yellowing; simultaneously, plant defense mechanisms are stimulated. To do this, genes involved in the phenylpropanoid pathway, phytohormone signaling, antioxidant production, and pathogenesis-related proteins are up- or downregulated, all according to cultivar susceptibility and plant organ. Furthermore, the displayed results bring out functional overlap within the same gene family among different plant species during plant–pathogen interactions. According to each time-point of stress development, differential gene expression between susceptible and resistant cultivars is a key point in breeding. Thanks to cheaper high-throughput sequencing technologies, transcriptome sequencing, and the ability to obtain reliable data, key resistance genes and knowledge of regulatory mechanisms can facilitate the development of breeding programs that represent a new level of defense for controlling biotic and abiotic stresses.

Author Contributions: Conceptualization: F.G. and A.M.; writing—original draft preparation: F.G. and A.M.; writing—review and editing: F.G., G.M., A.I. and A.M. All authors have read and agreed to the published version of the manuscript.

Funding: This study was carried out within the Agritech National Research Center and received funding from the European Union Next-GenerationEU (piano nazionale di ripresa e resilienza (pnrr)–missione 4 componente 2, investimento 1.4–D.D. 1032 17/06/2022, CN00000022).

Acknowledgments: During the preparation of this manuscript, the authors used Chat GPT (https://chatgpt.com, online version) to prepare Figure 1. The authors have reviewed and edited the output and take full responsibility for the content of this publication.

Conflicts of Interest: The authors declare no conflicts of interest.

List of Contributions:

1. Zhang, K.; Xing, W.; Sheng, S.; Yang, D.; Zhen, F.; Jiang, H.; Yan, C.; Jia, L. Genome-Wide Identification and Expression Analysis of Eggplant DIR Gene Family in Response to Biotic and Abiotic Stresses. *Horticulturae* **2022**, *8*, 732. https://doi.org/10.3390/horticulturae8080732.
2. Xiao, L.; Zhou, Z.; Zhu, C.; Zhao, J.; Hu, Z.; Liu, S.; Zhou, Y. Molecular Cloning, Characterization, and Expression Analysis of SIMILAR TO RCD-ONE (SRO) Family Genes Responding to Abiotic and Biotic Stress in Cucumber. *Horticulturae* **2022**, *8*, 634. https://doi.org/10.3390/horticulturae8070634.
3. Liu, H.; Chen, K.; Yang, L.; Han, X.; Wu, M.; Shen, Z. Physiological and Transcriptomic Analyses Reveal the Response of Medicinal Plant *Bletilla striata* (Thunb. ex A. Murray) Rchb. f. via Regulating Genes Involved in the ABA Signaling Pathway, Photosynthesis, and ROS Scavenging under Drought Stress. *Horticulturae* **2023**, *9*, 307. https://doi.org/10.3390/horticulturae9030307.
4. Zhou, P.; Huang, L.; Wang, Y.; Li, X.; Feng, X.; Li, L. Stepwise Optimization of the RT-qPCR Protocol and the Evaluation of Housekeeping Genes in Pears (*Pyrus bretschneideri*) under

Various Hormone Treatments and Stresses. *Horticulturae* **2023**, *9*, 275. https://doi.org/10.3390/horticulturae9020275.
5. Ambarwati, E.; Arwiyanto, T.; Widada, J.; Alam, T.; Andika, I.; Taryono. The Genes Associated with Jasmonic Acid and Salicylic Acid Are Induced in Tropical Chili Pepper against *Ralstonia solanacearum* by Applying Arbuscular Mycorrhizal Fungi. *Horticulturae* **2022**, *8*, 876. https://doi.org/10.3390/horticulturae8100876.
6. Nishijima, R.; Sanjaya, A.; Shinoyama, H.; Kazama, Y. Touch-Induced Transcriptional Changes in Flower Buds of a Non-Model Horticultural Plant *Dianthus hybrida*. *Horticulturae* **2022**, *8*, 918. https://doi.org/10.3390/horticulturae8100918.
7. Huang, X.; He, L.; Tan, H.; Liu, J.; Qiu, Q.; Sun, Q.; Ouyang, L.; Han, H.; He, Q. Transcriptome and Physiological Analyses of Resistant and Susceptible Pepper (*Capsicum annuum*) to *Verticillium dahliae* Inoculum. *Horticulturae* **2024**, *10*, 1160. https://doi.org/10.3390/horticulturae10111160.
8. Labarile, R.; Mincuzzi, A.; Spanò, R.; Mascia, T. Virus-Induced Silencing of a Sequence Coding for Loricrin-like Protein in *Phytophthora infestans* upon Infection of a Recombinant Vector Based on Tobacco Mosaic Virus. *Horticulturae* **2023**, *9*, 360. https://doi.org/10.3390/horticulturae9030360.
9. Liu, S.; Zhang, M.; Wang, H.; Xu, Y.; Wen, C.; Zhang, J.; Zhang, Y.; Shi, H. Coordinated Regulation of Iron-Acquisition Genes and Citrate Biosynthesis Drives Seasonal Iron Deficiency Adaptation in 'Yali' Pears (*Pyrus bretschneideri* Rehd.). *Horticulturae* **2025**, *11*, 460. https://doi.org/10.3390/horticulturae11050460.

References

1. Klepikova, A.V.; Penin, A.A. Gene expression maps in plants: Current state and prospects. *Plants* **2019**, *8*, 309. [CrossRef] [PubMed]
2. Kazan, K. Diverse roles of jasmonates and ethylene in abiotic stress tolerance. *Trends Plant Sci.* **2015**, *20*, 219–229. [CrossRef] [PubMed]
3. Ullah, A.; Manghwar, H.; Shaban, M.; Khan, A.H.; Akbar, A.; Ali, U.; Ali, E.; Fahad, S. Phytohormones enhanced drought tolerance in plants: A coping strategy. *Environ. Sci. Pollut. Res.* **2018**, *25*, 33103–33118. [CrossRef] [PubMed]
4. Bharath, P.; Gahir, S.; Raghavendra, A.S. Abscisic Acid-Induced Stomatal Closure: An Important Component of Plant Defense Against Abiotic and Biotic Stress. *Front. Plant Sci.* **2021**, *12*, 615114. [CrossRef] [PubMed]
5. Mincuzzi, A.; Ippolito, A.; Brighenti, V.; Marchetti, L.; Benvenuti, S.; Ligorio, A.; Pellati, F.; Sanzani, S.M. The effect of polyphenols on pomegranate fruit susceptibility to *Pilidiella granati* provides insights into disease tolerance mechanisms. *Molecules* **2020**, *25*, 515. [CrossRef] [PubMed]
6. Harbaugh, D.T.; Nepokroeff, M.; Rabeler, R.K.; McNeill, J.; Zimmer, E.A.; Wagner, W.L. A new lineage-based tribal classification of the family Caryophyllaceae. *Int. J. Plant Sci.* **2010**, *171*, 185–198. [CrossRef]
7. Wang, Y.; Chen, B.; Cheng, C.; Fu, B.; Qi, M.; Du, H.; Geng, S.; Zhang, X. Comparative transcriptomics analysis reveals the differences in transcription between resistant and susceptible pepper (*Capsicum annuum* L.) varieties in response to anthracnose. *Plants* **2024**, *13*, 527. [CrossRef] [PubMed]
8. Ao, N.; Zou, H.; Li, J.; Shao, H.; Kageyama, K.; Feng, W. First report of *Pythium aphanidermatum* and *Pythium myriotylum* causing root rot on chili pepper (*Capsicum annuum* L.) in Guizhou, China. *Crop Prot.* **2024**, *181*, 106704. [CrossRef]

horticulturae

Article

Genome-Wide Identification and Expression Analysis of Eggplant *DIR* Gene Family in Response to Biotic and Abiotic Stresses

Kaijing Zhang [1], Wujun Xing [1], Suao Sheng [1], Dekun Yang [1], Fengxian Zhen [1], Haikun Jiang [2,3], Congsheng Yan [2,3] and Li Jia [2,3,*]

1 College of Agriculture, Anhui Science and Technology University, Chuzhou 233100, China
2 Institute of Horticulture, Anhui Academy of Agricultural Sciences, Hefei 230001, China
3 Key Laboratory of Genetic Improvement and Ecophysiology of Horticultural Crop, Hefei 230001, China
* Correspondence: jiali820@aaas.org.cn

Abstract: Dirigent proteins (*DIR*) play important roles in the biosynthesis of lignins and lignans, defensive responses, secondary metabolism, and disease resistance in plants. The *DIR* gene family has been identified and studied in many plants. However, the identification of *DIR* gene family in eggplant has not been conducted yet. Therefore, in this study, based on the available genome information of eggplant, the *DIR* family genes in eggplant were identified with bioinformatics methods. The expression pattern analyses of eggplant *DIR* family genes in different organs and stresses were also conducted to understand their biological functions. The results showed that a total of 24 *DIR* genes were identified in the eggplant, which were divided into three subfamilies (*DIR-a*, *DIR-b/d*, and *DIR-e*). Synteny analysis of *DIR* genes among eggplant, *Arabidopsis*, and rice showed that 15 eggplant *DIR* genes were colinear with 18 *Arabidopsis DIR* genes, and 16 eggplant *DIR* genes were colinear with 15 rice *DIR* genes. Phylogenetic tree analysis showed that 19 pairs of orthologous genes were identified between eggplant and pepper. The *cis*-acting elements analysis implied that the eggplant *DIR* genes contained a lot of *cis*-elements associated with stress and hormone response. The organ-specific expression analysis of eggplant *DIR* family genes revealed that only the *SmDIR3* gene was highly expressed in all the 19 organs of eggplant. Some *SmDIR* genes, including *SmDIR7*, *SmDIR8*, *SmDIR11*, *SmDIR14*, *SmDIR18*, *SmDIR19*, *SmDIR20*, and *SmDIR23*, were not or were lowly expressed in the eggplant organs, while the other eggplant *DIR* family genes showed an organ-specific expression pattern. Furthermore, 19 of 24 *SmDIR* genes were differentially expressed in response to abiotic and biotic stresses. 5 *SmDIR* genes, including *SmDIR3*, *SmDIR5*, *SmDIR6*, *SmDIR12*, and *SmDIR22*, were differentially expressed under multiple types of abiotic and biotic stresses. Especially notable, the *SmDIR22* gene was differentially expressed under three types of abiotic stresses and two types of biotic stresses, which indicated that the *SmDIR22* gene plays an important role in the response to abiotic and biotic stresses. These results provide valuable evidence for a better understanding of the biological role of *DIR* genes in eggplant.

Keywords: eggplant; *DIR*; gene family; bioinformatics; expression analysis

1. Introduction

In nature, plants are easily influenced by pathogens, pests, severe temperatures, drought, salt, and heavy metals, which cause a dramatic reduction of crop yield and quality [1,2]. Plants usually respond to biotic and abiotic stresses by activating a variety of genes, including the dirigent protein (*DIR*) gene. The *DIR* gene is involved in the synthesis of lignins and lignans, which play a pivotal role against biotic and abiotic stresses in plants [3–5]. As an important compound, lignin confers stability and hydrophobicity to the plant vascular system and forms a barrier against microbial pathogens and pests [6,7]. Lignan has antifungal properties by inhibiting microbe-derived degradative enzymes such as cellulases, laccases, polygalacturonases, and glucosidases [8–10]. Besides, lignans can also defend against insect attacks by disrupting the insect endocrine system [11]. *DIR* genes can regulate the biosynthesis of lignins and lignans, which indicates that *DIR* genes actively respond to biotic and abiotic stresses.

DIR proteins are widely distributed in almost all vascular plants, including ferns, gymnosperms, and angiosperms [12–14]. Previous studies reported that *DIR* proteins are divided into five subgroups, including *DIR-a*, *DIR-b*, *DIR-c*, *DIR-d*, and *DIR-e* [3]. With the increasing number of *DIR* proteins, another two new subgroups, *DIR-f* and *DIR-g*, were generated, and the *DIR-b* and *DIR-d* subfamilies are combined together as the *DIR-b/d* subgroup [12]. *DIR* proteins were first identified in *Forsythia suspense* [15]. Subsequently, the identification of *DIR* family genes were reported in a variety of monocotyledon and dicotyledon plants, such as *Arabidopsis thaliana* (L.) Heynh. [16], rice (*Oryza sativa* L.) [17,18], spruce (*Picea* spp.) [3,12], pear (*Pyrus bretschneideri*) [5], *Populus* [19], Chinese cabbage (*Brassica rapa* L.) [4], pepper (*Capsicum annuum* L.) [20], strawberry (*Fragaria vesca*) [21], Cucurbitaceae [22], and so on. Moreover, many studiers reported that *DIR* family genes play important regulatory roles in the response to biotic and abiotic stresses, such as high temperature [13,18], low temperature [4,18], drought [4,13,18,23–25], salt [18,23,25,26], heavy metals [18], H_2O_2 [13,23], and diseases [3,4,25,27–31]. In addition, *DIR* genes also actively respond to hormone stresses including ethylene, gibberellin (JA), salicylic acid (SA), abscisic acid (ABA), and so on [4,5,13,25].

Eggplant (*Solanum melongena* L.), one of the most important solanaceous crops, is widely cultivated across the world for its fruits, ranking third for total production and economic value in Solanaceae species, after potato (*Solanum tuberosum* L.) and tomato (*Solanum lycopersicum* L.) [32,33]. As early as 2014, the first genome sequencing project of eggplant was completed [34]. With the rapid development of genome sequencing technology, the improved eggplant genomes were constantly updated [35–38]. By using the high-quality genome information of eggplant, a great deal of research on gene family identification has been performed, such as the identification of R2R3-MYB [39], WRKY [40], NAC [41], ARF [42], and other gene families. However, the genome-wide identification of *DIR* gene family in eggplant was still not reported, which greatly limits the functional studies of *DIR* genes in eggplant.

Therefore, based on the high-quality genome information of eggplant, *DIR* family genes in eggplant (*SmDIR* genes) were identified for the first time in this study. The physicochemical characteristics, chromosome locations, gene structures, conserved motifs, phylogenetic tree, *cis*-acting elements, and synteny of *SmDIR* genes were analyzed via bioinformatic methods. Moreover, the analyses of organ-specific expression and stress-responsive gene expression patterns of *SmDIR* genes were conducted to preliminarily explore the biological functions of *DIR* family genes in eggplant. The results of the current study will provide future insight for further research on the biological functions of *SmDIR* genes, as well as a theoretical reference for the resistance breeding of eggplant.

2. Materials and Methods

2.1. Identification and Chromosomal Mapping of DIR Family Genes in Eggplant

The Hidden Markov Model (HMM) of dirigent domain (PF03018) was used to identify the putative *DIR* family genes in eggplant genome V4.1 [38] using HMMER 3.0 software [43] with E-value $< 1 \times 10^{-5}$. The protein sequences of candidate *SmDIR* genes were extracted with Fasta Extract program in TBtools software [44]. The Pfam (http://pfam.xfam.org/, accessed on 26 May 2022) [45] and SMART (http://smart.embl.de/smart/batch.pl, accessed on 26 May 2022) [46] websites were used to confirm the conserved dirigent domain. Genes containing the whole dirigent domain were chosen as the final confirmed members of *DIR* gene family in eggplant. The chromosomal locations of *SmDIR* genes were analyzed and visualized with TBtools software. The physicochemical characteristics such as the number of amino acids, molecular weight, theoretical isoelectric point, instability index, aliphatic index, and grand average of hydropathicity of each *SmDIR* protein were calculated with the online tool ExPASy (https://web.expasy.org/protparam/, accessed on 27 May 2022). The subcellular localization of *SmDIR* genes were predicted with the online website CELLO (http://cello.life.nctu.edu.tw/, accessed on 27 May 2022) [47].

2.2. Gene Structure Analysis and Phylogenetic Tree Construction of DIR Family Genes in Eggplant

The exon/intron structures of *SmDIR* genes were drawn with GFF3 file (General Feature Format 3) of eggplant genome V4.1 using TBtools software. The conserved motifs of *SmDIR* proteins were analyzed using the MEME software (version = 5.4.1) [48] with the following parameters: maximum number of misfits, 10; and the optimum width of each motif, 6–100 amino acids. The 1.5 kb upstream sequences of the translation initiation codon of *SmDIR* genes were extracted from the eggplant genome V4.1. The sequences were then submitted to PlantCARE (http://bioinformatics.psb.ugent.be/webtools/plantcare/html/, accessed on 27 May 2022) website [49] for *cis*-acting elements prediction. The phylogenetic tree of *DIR* family genes in eggplant, pepper [20] and *Arabidopsis* [16] was generated using MEGA 11 software [50] with the neighbor-joining method and 1000 bootstrap replications. The final phylogenetic tree was visualized using the online website Evolview (https://www.evolgenius.info/evolview-v3/#login, accessed on 27 May 2022) [51].

2.3. Gene Duplication and Synteny Analysis of DIR Family Genes

The tandem and segmental duplications of *SmDIR* genes were analyzed with MCScanX software [52]. The synteny analysis of *DIR* family genes among eggplant, *Arabidopsis* and rice was conducted with MCScanX software. The syntenic relationships of *DIR* family genes among eggplant, *Arabidopsis*, and rice were visualized with Circos software [53].

2.4. RNA-Seq Reanalysis of Eggplant Transcriptome Sequencing Data

The published eggplant transcriptome sequencing data were downloaded from the SRA database using the corresponding accession numbers (https://www.ncbi.nlm.nih.gov/sra, accessed on 4 June 2022). The downloaded SRA data were converted into Fastq data by using fasterq-dump.2.11.0 tool (https://github.com/ncbi/sra-tools/wiki/HowTo:-fasterq-dump, accessed on 10 June 2022). The read quality of Fastq data was then evaluated using the FastQC software (http://www.bioinformatics.babraham.ac.uk/projects/fastqc/, accessed on 10 June 2022). The adapter sequences and low-quality bases were removed from the Fastq data with Trimmomatic software (version = 0.39) [54], finally obtaining the filtered clean data. The index of eggplant genome V4.1 was built using STAR software (version = 2.7.10a) [55]. The filtered clean data were mapped to eggplant genome V4.1, generating the SAM files. Next, the SAM files were directly converted into the sorted BAM files using the SAMtools software (version = 1.15) [56]. The expression levels of each gene were estimated with StringTie software (v2.2.1) [57]. Finally, the differential expression analysis was performed using the count matrix in DESeq2 software [58].

2.5. Organ-Specific Expression Analysis of DIR Family Genes in Eggplant

The transcriptome sequencing data of 19 kinds of eggplant organs, including cotyledon, radicle, root, stem, flower, pistil, leaf, bud, fruit peduncle, fruit skin, fruit flesh, fruit, and so on, were retrieved and downloaded from the SRA database (PRJNA328564) [35]. The organ-specific expression analysis of eggplant *DIR* family genes was conducted using the above methods. The heatmap was drawn using the Heatmap program in TBtools software.

2.6. The Expression Profiles Analysis of Eggplant DIR Family Genes under Stresses

The transcriptome sequencing data of eggplant under stress treatments including high temperature (PRJNA531285) [59], low temperature (PRJNA572318) [60], salt (PRJNA477924) [61], bacterial wilt (caused by *Ralstonia solanacearum*, PRJNA728497) [62], *Verticillium* wilt (PRJNA451240) [63], and *Tuta absoluta* (PRJNA695350) [64] were retrieved and downloaded from the SRA database. The expression profiles of eggplant *DIR* family genes were analyzed using the above methods, and then drawn into heatmaps using TBtools software.

3. Results

3.1. Genome-Wide Identification of DIR Family Genes in Eggplant

A total of 24 *DIR* family genes were identified in the eggplant genome (Table 1). These eggplant *DIR* family genes were renamed from *SmDIR1* to *SmDIR24* according to their chromosomal locations. The detailed information of eggplant *DIR* family genes was summarized in Table 1. The 24 *SmDIR* proteins ranged from 146 (*SmDIR11*) to 546 (*SmDIR17*) amino acid (aa) in length, and the corresponding molecular weights varied from 16.14 kDa (*SmDIR11*) to 59.69 kDa (*SmDIR17*). The isoelectric points (pI) of the 24 *SmDIR* proteins ranged from 4.34 (*SmDIR9*) to 9.85 (*SmDIR19*). Among the 24 eggplant *DIR* proteins, 14 *SmDIR* proteins were basic proteins (pI > 7), and 10 *SmDIR* proteins were acidic proteins (pI < 7). 20 *SmDIR* proteins were stable proteins (instability index less than 40), and the other 4 *SmDIR* proteins (*SmDIR1*, *SmDIR8*, *SmDIR9*, and *SmDIR20*) were unstable proteins (instability index greater than 40). The aliphatic indexes of 24 *SmDIR* proteins varied from 66.53 (*SmDIR16*) to 101.55 (*SmDIR22*). Most of *SmDIR* proteins (79%) were hydrophobic proteins (grand average of hydropathicity > 0), and only five *SmDIR* proteins (*SmDIR1*, *SmDIR3*, *SmDIR8*, *SmDIR16*, and *SmDIR20*) were hydrophilic proteins (grand average of hydropathicity < 0). Subcellular localization prediction showed that most of the *SmDIR* genes were located in the plasma membrane, and the other *DIR* family members were located in the extracellular space and mitochondrion. (Table 1).

Based on the GFF3 file of eggplant genome V4.1, the chromosomal distributions of 24 *SmDIR* genes were mapped across the eggplant chromosomes using TBtools software (Figure 1). In the eggplant genome, the 24 *SmDIR* family genes were unevenly distributed on 10 of 12 eggplant chromosomes. There were no *SmDIR* genes located on chromosomes 3 and 10. Chromosome 12 harbored the largest number of 4 *SmDIR* genes, whereas chromosomes 4, 7, and 9 contained only one *SmDIR* gene, respectively. The chromosomes 1, 2, 5, 6, and 11 had three *SmDIR* genes, respectively. Two *SmDIR* genes were positioned on chromosome 8. In our study, five pairs of tandem duplication genes on chromosome 1 (*SmDIR1/SmDIR2*), chromosome 5 (*SmDIR8/SmDIR9*), chromosome 8 (*SmDIR15/SmDIR16*), and chromosome 12 (*SmDIR21/SmDIR22*, *SmDIR22/SmDIR23*) were identified in eggplant *DIR* gene family.

3.2. Phylogenetic Analysis of DIR Family Genes in Eggplant, Pepper and Arabidopsis

To investigate the evolutionary relationship of *DIR* family genes in eggplant, a phylogenetic tree was constructed with 24 *SmDIR* proteins, 24 pepper *DIR* proteins [20] and 25 *Arabidopsis DIR* proteins [16] by multiple sequence alignment (Figure 2). According to the clustering result of *Arabidopsis DIR* family genes, the phylogenetic tree was divided into three subgroups, including *DIR-a*, *DIR-b/d*, and *DIR-e*. Among them, the *DIR-b/d* subgroup harbored the largest number of *DIR* genes, containing a total of 40 *DIR* genes. The *DIR-e* subgroup had the second number of *DIR* genes, containing 19 *DIR* genes. The *DIR-a* subgroup had the least number of *DIR* genes, containing 14 *DIR* genes. In the phylogenetic tree, 19 pairs of orthologous genes were identified between eggplant and pepper, which were *SmDIR1/CaDIR17*, *SmDIR2/CaDIR16*, *SmDIR3/CaDIR5*, *SmDIR4/CaDIR8*, *SmDIR5/CaDIR13*, *SmDIR6/CaDIR6*, *SmDIR8/CaDIR10*, *SmDIR9/CaDIR19*, *SmDIR10/CaDIR23*, *SmDIR12/CaDIR22*, *SmDIR13/CaDIR12*, *SmDIR14/CaDIR14*, *SmDIR15/CaDIR2*, *SmDIR16/CaDIR1*, *SmDIR17/CaDIR7*, *SmDIR18/CaDIR4*, *SmDIR19/CaDIR9*, *SmDIR22/CaDIR24*, and *SmDIR24/CaDIR20*, respectively. The homologous genes clustered together in the evolutionary relationship possessed similar gene functions. Thus, the biological functions of *SmDIR* genes could be inferred from the molecular functions of pepper *DIR* family genes based on the clustering results of *DIR* family genes between eggplant and pepper.

Table 1. The physiochemical characteristics of *DIR* family genes in eggplant.

Gene Name	Gene ID	Number of Amino Acid (aa)	Molecular Weight (kDa)	Theoretical pI	Instability Index	Aliphatic Index	Grand Average of Hydropathicity	Subcellular Location
SmDIR1	*SMEL4.1_01g003540.1*	156	16.97	9.36	45.55	76.92	−0.108	Mitochondrion
SmDIR2	*SMEL4.1_01g003550.1*	343	36.23	6.13	11.54	95.54	0.171	Plasma Membrane
SmDIR3	*SMEL4.1_01g031140.1*	191	21.33	9.57	28.14	79.69	−0.017	Plasma Membrane
SmDIR4	*SMEL4.1_02g003080.1*	190	21.06	9.67	24.42	94.95	0.065	Mitochondrion
SmDIR5	*SMEL4.1_02g009260.1*	186	20.69	7.06	36.88	91.24	0.196	Plasma Membrane
SmDIR6	*SMEL4.1_02g018810.1*	194	21.19	9.08	28.43	93.87	0.111	Extracellular Space
SmDIR7	*SMEL4.1_04g004510.1*	190	21.37	9.42	29.75	91.79	0.022	Plasma Membrane
SmDIR8	*SMEL4.1_05g003580.1*	244	27.03	5.90	43.08	79.88	−0.167	Plasma Membrane
SmDIR9	*SMEL4.1_05g003590.1*	438	44.86	4.34	40.80	87.99	0.075	Extracellular Space
SmDIR10	*SMEL4.1_05g019870.1*	249	26.06	5.46	36.58	85.42	0.115	Plasma Membrane
SmDIR11	*SMEL4.1_06g013950.1*	146	16.14	8.82	27.09	91.51	0.135	Plasma Membrane
SmDIR12	*SMEL4.1_06g020190.1*	212	23.63	9.73	29.47	89.62	0.094	Plasma Membrane
SmDIR13	*SMEL4.1_06g027190.1*	372	38.41	4.75	26.34	96.02	0.230	Plasma Membrane
SmDIR14	*SMEL4.1_07g016780.1*	177	19.52	5.22	27.96	95.31	0.124	Plasma Membrane
SmDIR15	*SMEL4.1_08g026590.1*	190	21.28	7.79	13.38	88.84	0.088	Plasma Membrane
SmDIR16	*SMEL4.1_08g026600.1*	167	19.13	6.02	26.49	66.53	−0.298	Plasma Membrane
SmDIR17	*SMEL4.1_09g022870.1*	546	59.69	8.52	38.73	97.33	0.354	Plasma Membrane
SmDIR18	*SMEL4.1_11g000960.1*	168	18.76	6.90	15.68	99.76	0.123	Plasma Membrane
SmDIR19	*SMEL4.1_11g019830.1*	178	19.93	9.85	35.35	100.34	0.069	Plasma Membrane
SmDIR20	*SMEL4.1_11g026460.1*	369	38.79	4.63	42.99	84.80	−0.063	Extracellular Space
SmDIR21	*SMEL4.1_12g003420.1*	184	20.19	9.52	26.98	90.11	0.123	Mitochondrion
SmDIR22	*SMEL4.1_12g003430.1*	194	20.87	9.24	22.53	101.55	0.198	Extracellular Space
SmDIR23	*SMEL4.1_12g003440.1*	193	21.21	6.96	23.74	93.37	0.130	Extracellular Space
SmDIR24	*SMEL4.1_12g003620.1*	194	21.36	7.87	19.27	79.85	0.055	Extracellular Space

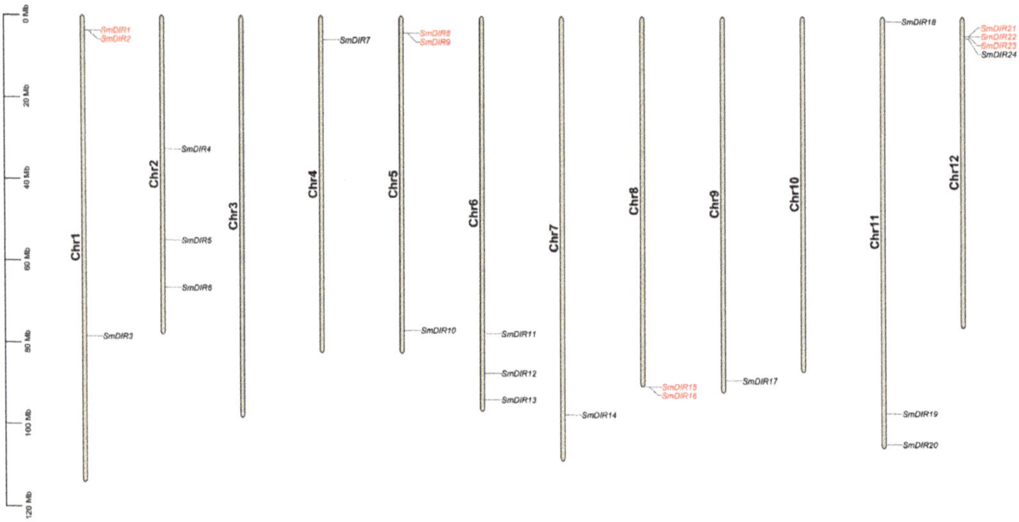

Figure 1. The chromosomal locations of *DIR* family genes in eggplant.

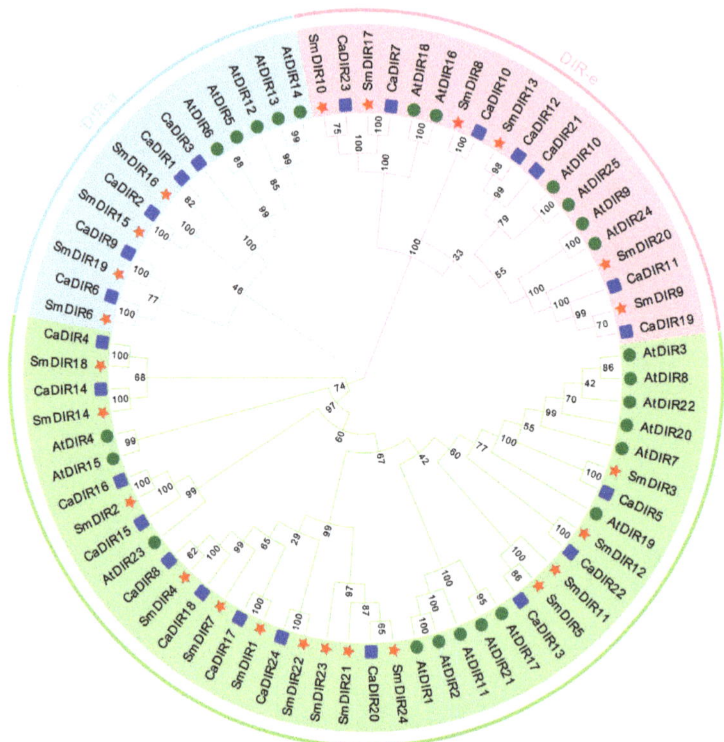

Figure 2. Phylogenetic analysis of *DIR* proteins from eggplant, pepper and *Arabidopsis*.

3.3. Gene Structure and Conserved Motifs Analyses of Eggplant DIR Genes

The diagrams of phylogenetic tree and gene structures of *SmDIR* genes were drawn with TBtools software. The results of phylogenetic tree analysis showed that 24 *SmDIR* genes were classified into three subgroups, namely *DIR-a*, *DIR-b/d* and *DIR-e* (Figure 3), which was consistent with the clustering result of eggplant, pepper and *Arabidopsis DIR* genes (Figure 2). Among them, there were fourteen *DIR* genes in the *DIR-b/d* subgroup, six *DIR* genes in the *DIR-e* subgroup, and four *DIR* genes in the *DIR-a* subgroup. The gene structure analysis showed that most of *DIR* genes in subgroup *DIR-b/d* contained one exon. All four *DIR* genes in the *DIR-a* subgroup contained only one exon. All *DIR* genes except *SmDIR10*, *SmDIR13*, and *SmDIR17* in subgroup *DIR-e* contained one exon (Figure 3). The results showed that most of the *SmDIR* genes (71%) contained only one exon and no introns, suggesting that the exon number of *SmDIR* genes was relatively conservative.

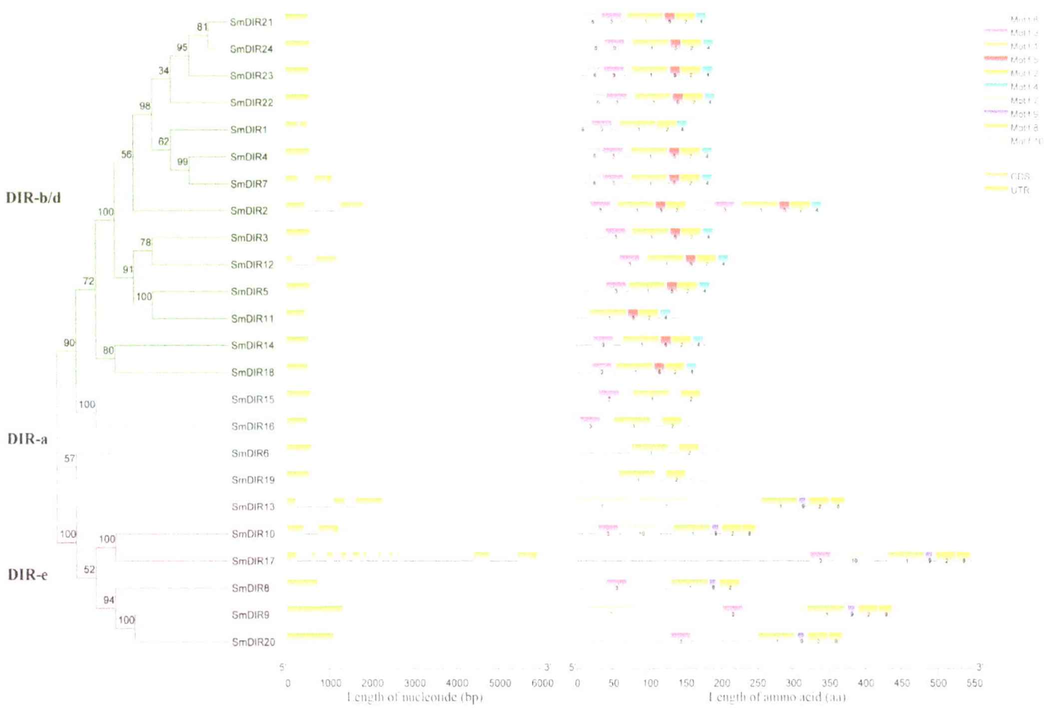

Figure 3. Exon-intron structures of *SmDIR* genes and conserved motifs of *SmDIR* proteins.

A schematic diagram of conserved motifs of *SmDIR* proteins was constructed with MEME analysis (Figure 3). A total of ten motifs were revealed in the eggplant *DIR* proteins (Table 2). The motif compositions of *SmDIR* proteins in different subgroups were different, while *SmDIR* proteins in the same subgroup shared the similar number, type, and order of motifs. For example, most of *SmDIR* proteins in the subgroup *DIR-b/d* contained motifs 3, 1, 5, 2, and 4, and exhibited the same order. In the subgroup *DIR-a*, motifs 1 and 2 were found in all *SmDIR* proteins and showed the same order. In the subgroup *DIR-e*, motifs 1, 9, 2, and 8 were found in all *SmDIR* proteins and showed the same order. These results indicated that the different distribution of conserved motifs in different subgroups might lead to the evolution of *SmDIR* gene functional diversity. The similar conserved motifs of *SmDIR* proteins in the same subgroup indicated that they have similar functions.

Table 2. The motifs information of *DIR* proteins in eggplant.

Motif	Sequence	Number of Amino Acid	Pfam Annotation
motif 1	FGTLTVIDDPLTIGPEPNSKJIGRAQGIYGSASQNGVSLLMALNFVFTGGKY	52	Dirigent
motif 2	YREMAIVGGTGKFRLARGYATAKTYWFB	28	Dirigent
motif 3	HAKEKVTKLHFYFHDILSGKNPTAIQIAQ	29	-
motif 4	TGDAIVEYNVVVLHY	15	-
motif 5	GSTLSILGRNPVFHE	15	-
motif 6	MPTAQGIDLGPKAVEKWFKKL	21	-
motif 7	MAIIQNSNSFQLKVIISYLLILALTITFATAGRILDEEVATPTVPPNNPDPTDQPPVSGATAAGAAAGAT	70	-
motif 8	GQHTTDGVETILHITVYJTY	20	-
motif 9	DSLSFFGVHR	10	-
motif 10	IYSGQIPFATPLGFQPPEDGVAIPNANGAMPTFNINGVPLGTGLAGTIFAGGNN	54	-

3.4. The Synteny Analysis of DIR Genes among Eggplant, Arabidopsis and Rice

To better understand the molecular functions of eggplant *DIR* family genes, the synteny analysis of *DIR* family genes among eggplant, *Arabidopsis* and rice was conducted. The results showed 29 syntenic relationships between 15 *SmDIR* genes (*SmDIR2, SmDIR3, SmDIR6, SmDIR9, SmDIR10, SmDIR11, SmDIR12, SmDIR13, SmDIR15, SmDIR17, SmDIR19, SmDIR20, SmDIR21, SmDIR23, SmDIR24*) genes and 18 *Arabidopsis DIR* genes (*AT4G13580, AT3G24020, AT1G58170, AT1G55210, AT2G39430, AT3G13650, AT3G55230, AT2G28670, AT5G49040, AT2G21100, AT1G65870, AT1G64160, AT3G58090, AT5G42500, AT4G23690, AT3G13662, AT5G42655, AT4G11430*). There were 27 syntenic relationships between 16 *SmDIR* genes (*SmDIR2, SmDIR3, SmDIR4, SmDIR5, SmDIR6, SmDIR8, SmDIR9, SmDIR10, SmDIR12, SmDIR13, SmDIR15, SmDIR16, SmDIR17, SmDIR21, SmDIR23, SmDIR24*) and 15 rice *DIR* genes (*Os04g42720, Os03g05030, Os03g17220, Os01g06250, Os01g65700, Os11g07830, Os12g07580, Os11g27620, Os11g07690, Os03g59440, Os07g44930, Os07g44250, Os07g44280, Os01g62030, Os07g44450*). It was found that five *SmDIR* genes (*SmDIR1, SmDIR7, SmDIR14, SmDIR18, SmDIR22*) were not colinear with either *A. thaliana* or rice, suggesting that these five *SmDIR* genes were conservative in eggplant. The synteny analysis of *DIR* family genes in eggplant showed that there were four pairs of *SmDIR* genes (*SmDIR1/SmDIR21, SmDIR1/SmDIR24, SmDIR8/SmDIR13, SmDIR10/SmDIR17*) with syntenic relationships, which were segmental duplication gene pairs (Figure 4).

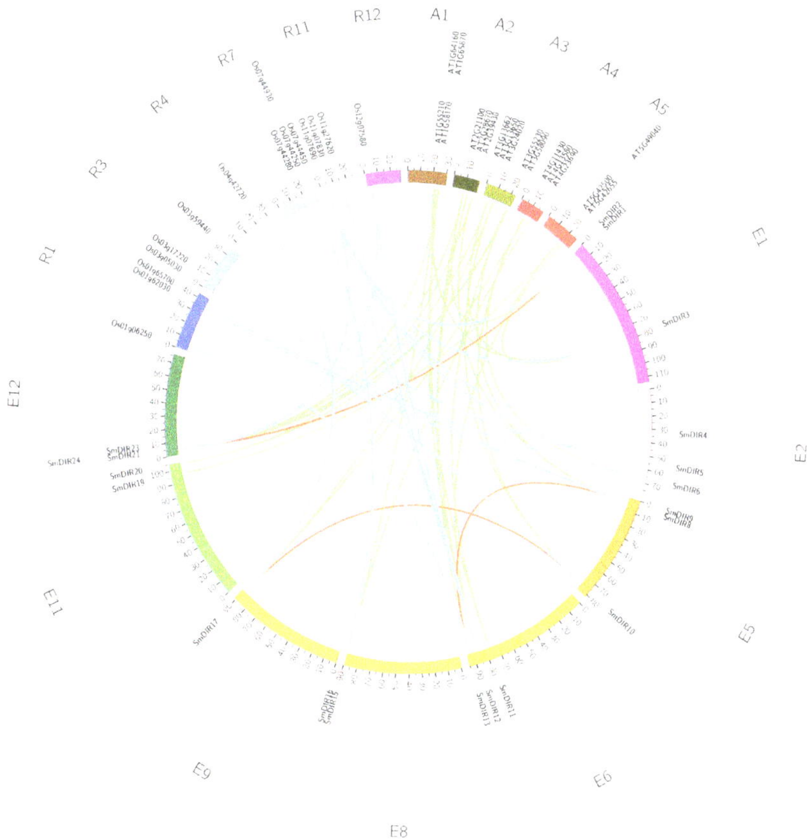

Figure 4. Syntenic relationships of *DIR* family genes among eggplant, *Arabidopsis* and rice.

3.5. The Cis-Acting Elements Analysis of Eggplant DIR Genes

The 1.5-kb upstream sequences from the transcription start site of *SmDIR* genes were extracted to analyze their *cis*-acting elements in the promoters. The results showed that 14 kinds of *cis*-elements were identified (Figure 5). Among them, *cis*-elements related to light responsiveness were the major type, including ACE, AE-box, ATCT-motif, Box 4, Box II, G-box, GT1-motif, I-box, LAMP-element, and so on, accounting for 31% of total *cis*-elements. In addition, some other *cis*-elements were also identified, including *cis*-elements related to hormone response (abscisic acid, MeJA, auxin, gibberellin, salicylic acid), stress response (drought, low temperature, defense), circadian control, endosperm expression, and meristem expression. The results indicated that eggplant *DIR* genes play vital roles in plant growth and development.

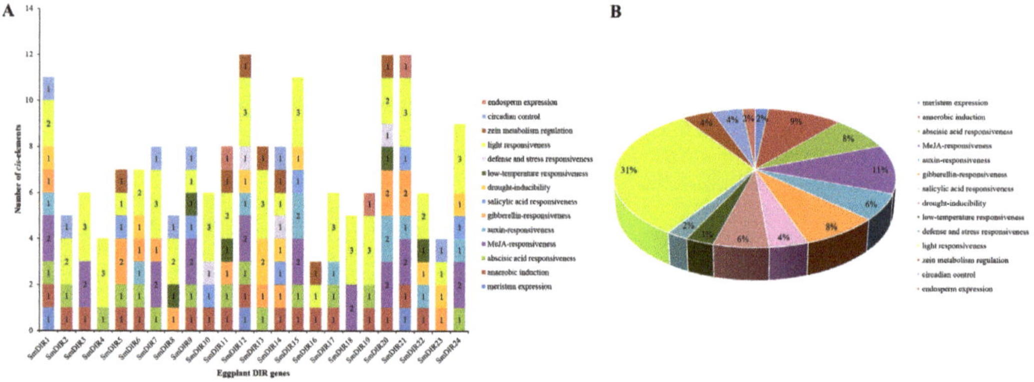

Figure 5. *Cis*-elements analysis of the promoters of eggplant *DIR* family genes. (**A**) The types and numbers of various *cis*-elements in the promoters of each *SmDIR* gene. (**B**) The relative proportions of different kinds of *cis*-elements in the promoters of *SmDIR* genes are displayed by pie chart.

3.6. Organ-Specific Expression Analysis of Eggplant DIR Genes

To analyze the organ-specific expression profiles of *SmDIR* genes in eggplant, the transcriptome sequencing data of 19 kinds of eggplant organs (accession number: PRJNA328564) were reanalyzed with eggplant genome V4.1. The results indicated that the eggplant *DIR* gene *SmDIR3* was highly expressed in all the 19 organs of eggplant. Some *SmDIR* genes, including *SmDIR7, SmDIR8, SmDIR11, SmDIR14, SmDIR18, SmDIR19, SmDIR20,* and *SmDIR23,* were not or were lowly expressed in all the eggplant organs. *SmDIR12* gene was expressed in all eggplant organs, with the lower expression levels in senescent leaf, fruit peduncle, fruit flesh stage 3, and fruit skin stage 3. The *SmDIR22* gene was expressed in all eggplant organs, with the highest expression level in root and the lowest expression levels in fruit flesh and skin at stage 3. The *SmDIR6* gene was highly expressed in fruit skin stage 3, stem and fruit peduncle, but was lowly expressed in cotyledon and fruit. Some *SmDIR* gene was organ-specific expressed. For example, the *SmDIR1* gene was highly expressed in bud, but was lowly or not expressed in the other eggplant organs. *SmDIR13* gene was highly expressed in radicle, root and bud, but was lowly or not expressed in the other eggplant organs. The eggplant *DIR* genes *SmDIR2, SmDIR4, SmDIR9, SmDIR10, SmDIR17,* and *SmDIR21* were highly expressed in radicle and root, but were lowly expressed in the other organs (Figure 6). In summary, the expression patterns of *SmDIR* genes varied in different organs, suggesting a degree of organ specificity.

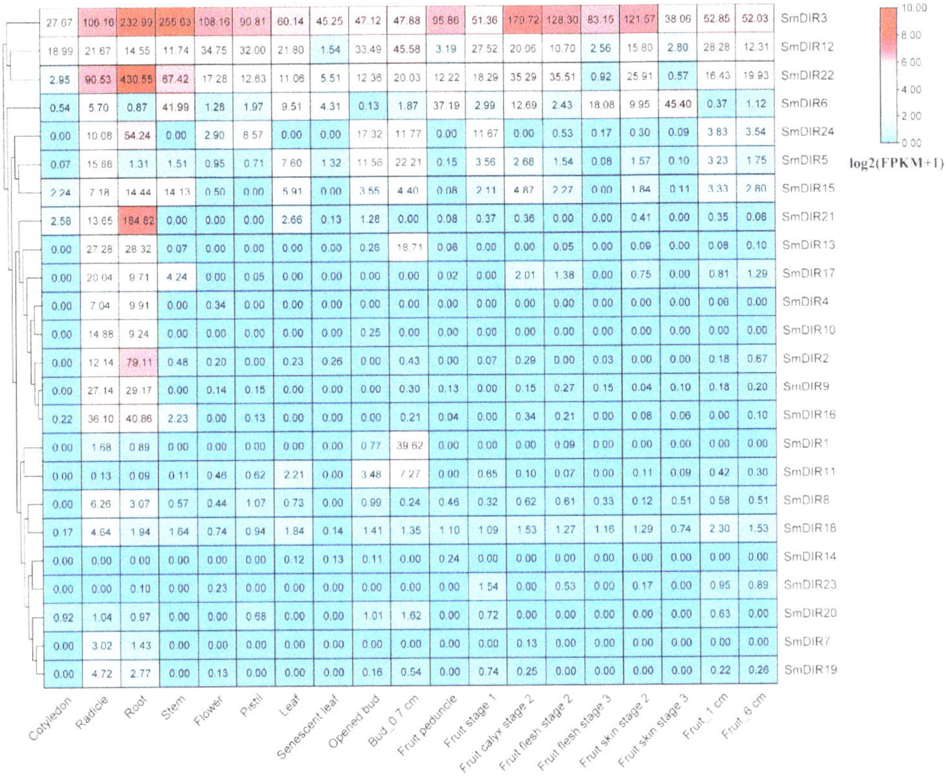

Figure 6. The expression heatmap of eggplant *DIR* family gene in different organs. The data in the boxes indicated the original FPKM values.

3.7. The Expression Profiles of Eggplant DIR Genes under Abiotic Stresses

To determine the expression profiles of eggplant *DIR* family genes under high temperature, low temperature, and salt stresses, the published transcriptome sequencing data (PRJNA531285, PRJNA572318, PRJNA477924) were re-analyzed with eggplant genome V4.1. Compared with the control, four eggplant *DIR* genes including *SmDIR3*, *SmDIR6*, *SmDIR12*, and *SmDIR22* genes were significantly down-regulated under high temperature stress. However, the expression patterns of these four *SmDIR* genes were different. Among them, the *SmDIR3* gene was significantly down-regulated only at 20DAF (days after flowering). *SmDIR12* gene was significantly down-regulated only at 10DAF. *SmDIR6* gene was significantly down-regulated at 15DAF and 20DAF. *SmDIR22* gene was significantly down-regulated at 10DAF, 15DAF, and 20DAF. The other 20 *SmDIR* genes were lowly or not expressed in the peel, and not affected by high temperature treatment (Figure 7A). Under low temperature stress, only two eggplant *DIR* genes *SmDIR5* and *SmDIR22* were significantly down-regulated in both sensitive and tolerant eggplant cultivars. Among the other twenty-two *SmDIR* genes, seventeen *SmDIR* genes were not or lowly expressed in the leaf, five *SmDIR* genes were not significantly regulated by low temperature treatment (Figure 7B). Under salt stress, only *SmDIR22* genes were significantly down-regulated in the sensitive eggplant leaf. The other 23 *SmDIR* genes were not significantly affected by salt stress (Figure 7C).

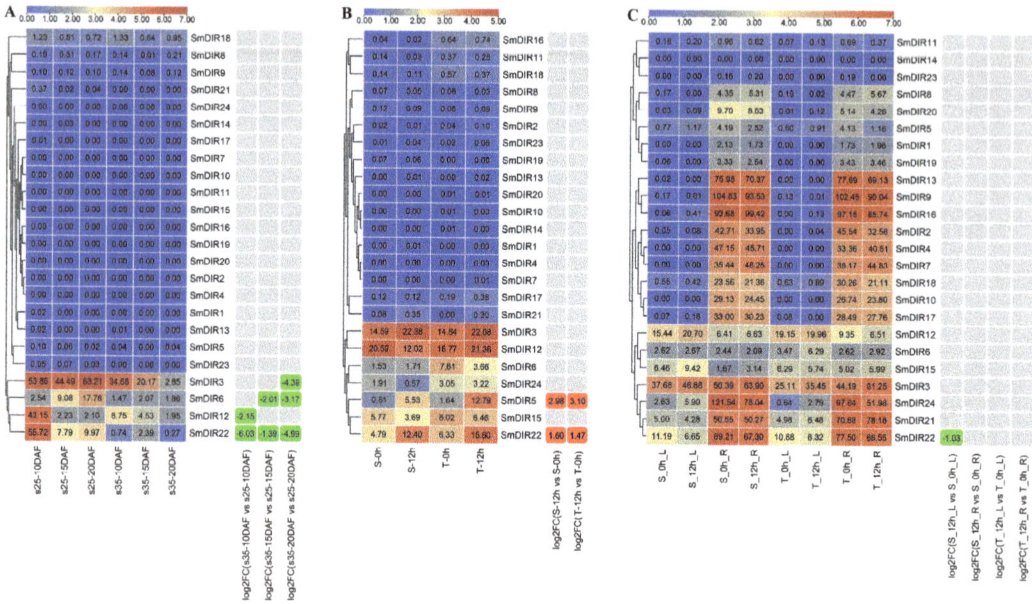

Figure 7. The expression heatmaps of eggplant *DIR* family genes under abiotic stress treatments. (**A**) The expression patterns of *SmDIR* genes under high temperature stress. s25: control treatment (25 °C); s35: high temperature treatment (35 °C); DAF: days after flowering. (**B**) The expression patterns of *SmDIR* genes under low temperature stress. S: low temperature-sensitive eggplant cultivar; T: low temperature-tolerant eggplant cultivar. 0 h: low temperature treatment for 0 h (control treatment); 6 h: low temperature treatment (4 °C) for 6 h. (**C**) The expression patterns of *SmDIR* genes under salt stress. S: salt-sensitive eggplant cultivar; T: salt-tolerant eggplant cultivar. 0 h: salt treatment for 0 h (control treatment); 12 h: salt treatment (200 mM NaCl) for 12 h; L: leaf; R: root. In each figure, the data in the boxes of left heatmap indicated the original FPKM values. Differentially expressed genes were highlighted by red (up-regulation) and green (down-regulation) color with log2 (fold-change) values in the right table.

3.8. The Expression Profiles of Eggplant DIR Genes under Biotic Stresses

Based on the eggplant genome V4.1, the published transcriptome sequencing data under bacterial wilt (PRJNA728497), *Verticillium* wilt (PRJNA451240), and *Tuta absoluta* stresses (PRJNA695350) were re-analyzed to investigate the expression profiles of eggplant *DIR* genes under biotic stresses. Under bacterial wilt stress, only five *SmDIR* genes were not affected, which were also not or lowly expressed in the samples. The other 19 *SmDIR* genes exhibited significantly differential expression under bacterial wilt infection. Among them, three *SmDIR* genes, including *SmDIR6*, *SmDIR12*, and *SmDIR21*, were up-regulated in both resistant and susceptible eggplant cultivars. *SmDIR6* and *SmDIR12* genes were up-regulated in both root and stem, while *SmDIR21* was only up-regulated in the root. Thirteen *SmDIR* genes were down-regulated in response to bacterial wilt infection. Whereas the expression patterns of these 13 *SmDIR* genes were different. Only three of these 13 *SmDIR* genes, including *SmDIR2*, *SmDIR3*, and *SmDIR18*, were down-regulated in the susceptible eggplant cultivar. *SmDIR3* was down-regulated in the root and stem of the susceptible eggplant cultivar. *SmDIR2* and *SmDIR18* were only down-regulated in the root of susceptible eggplant cultivar. Only one *SmDIR* gene *SmDIR8* was down-regulated in the root of resistant eggplant cultivar. Two *SmDIR* genes including *SmDIR16* and *SmDIR17* were down-regulated in both root and stem in susceptible and resistant eggplant cultivars. Six *SmDIR* genes, including *SmDIR4*, *SmDIR9*, *SmDIR10*, *SmDIR13*, *SmDIR19*, and *SmDIR24*, were down-regulated in the root of both susceptible and resistant eggplant cultivars. The *SmDIR5* gene was down-regulated in the root and stem of the susceptible eggplant line, but down-regulated only in the root of the resistant eggplant line. The expression patterns of three *SmDIR* genes (*SmDIR7*, *SmDIR15* and *SmDIR22*) were different in different samples, containing both up-regulation and down-regulation (Figure 8A).

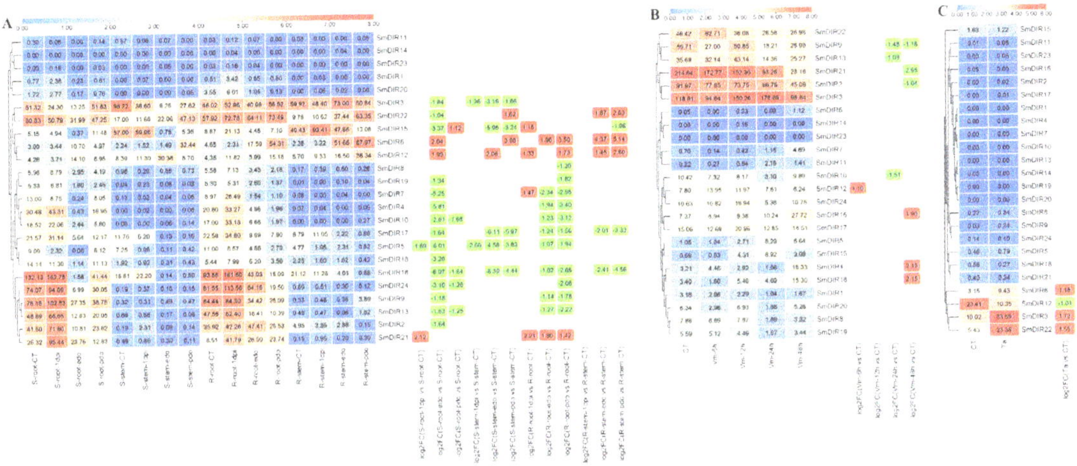

Figure 8. The expression heatmaps of eggplant *DIR* family genes under biotic stress treatments. (**A**) The expression patterns of *SmDIR* genes under bacterial wilt stress. S: susceptible eggplant cultivar; R: resistant eggplant cultivar; CT: control treatment; 1 dpi: 1 day post-inoculation; edo: early disease onset stage; pdo: peak disease onset stage. (**B**) The expression patterns of *SmDIR* genes under *Verticillium* wilt stress. CT: control treatment; Vm-6 h, Vm-12 h, Vm-24 h and Vm-48 h represented 6, 12, 24 and 48 h after *Verticillium* wilt infection. (**C**) The expression patterns of *SmDIR* genes under *Tuta absoluta* stress. CT: control treatment; Ta: *Tuta absoluta* infection. In each figure, the data in the boxes of the left heatmap indicated the original FPKM values. Differentially expressed genes were highlighted by red (up-regulation) and green (down-regulation) color with log2 (fold-change) values in the right table.

Under *Verticillium* wilt stress, a total of nine *SmDIR* genes exhibited significantly differential expression compared to the control treatment. Among them, five *SmDIR* genes, including *SmDIR2*, *SmDIR9*, *SmDIR10*, *SmDIR13*, and *SmDIR21*, were down-regulated at 24 or 48 h after *Verticillium* wilt infection, while four *SmDIR* genes, including *SmDIR4*, *SmDIR12*, *SmDIR16*, and *SmDIR18*, exhibited significantly up-regulated expression after *Verticillium* wilt infection. *SmDIR4*, *SmDIR16*, and *SmDIR12* genes were significantly up-regulated at 48 h after *Verticillium* wilt infection. *SmDIR18* genes were significantly up-regulated at 6 h after *Verticillium* wilt infection (Figure 8B). Under *Tuta absoluta* stress, only four *SmDIR* genes were significantly differentially expressed compared to the control. Among them, three *SmDIR* genes, including *SmDIR3*, *SmDIR6*, and *SmDIR22*, were up-regulated expression, one *SmDIR* gene *SmDIR12* was down-regulated expression after *Tuta absoluta* infection (Figure 8C).

3.9. The Regulation Patterns Analysis of Eggplant DIR Family Genes under Stresses

To analyze the expression patterns of *SmDIR* genes under biotic and abiotic stresses, the differentially expressed genes were labeled and drawn into a heatmap (Figure 9). The results showed that 19 of 24 *SmDIR* genes were differentially expressed in response to stresses. Among them, five *SmDIR* genes, including *SmDIR3*, *SmDIR5*, *SmDIR6*, *SmDIR12*, and *SmDIR22*, were differentially expressed under multiple types of abiotic and biotic stresses, indicating that these five *SmDIR* genes actively participate in stress responses. Especially notable, the *SmDIR22* gene was differentially expressed under three types of abiotic stresses and two types of biotic stresses, which could be considered as the favorable gene for the further studies. The other 14 of 24 *SmDIR* genes were only differentially expressed under biotic stresses. The regulation patterns analysis of eggplant *DIR* family genes could provide the theoretical references for further research on the biological function of eggplant *DIR* genes and the resistance breeding of eggplant.

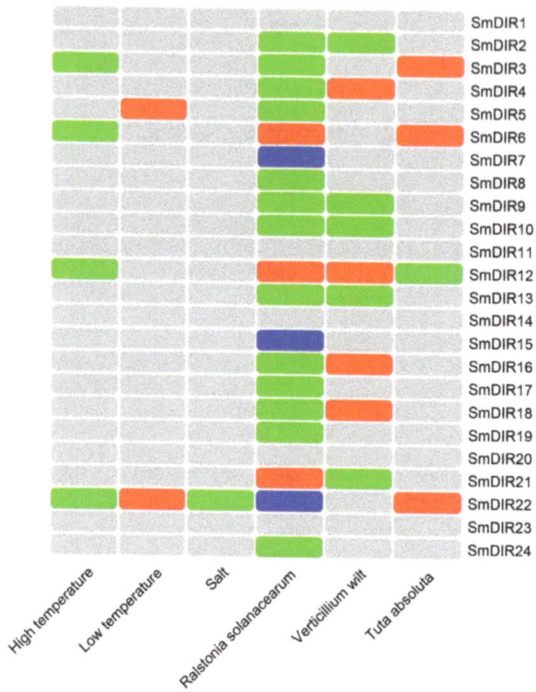

Figure 9. The expression patterns heatmap of eggplant *DIR* family genes under abiotic and biotic stresses. Gray color represents that the *SmDIR* genes were not differentially expressed; red color represents up-regulated expression; green color represents down-regulated expression; blue color represents that the *SmDIR* genes exhibited both up-regulated and down-regulated expressions.

4. Discussion

Since the *Arabidopsis thaliana* genome was published in 2000, more and more plant genomes have been sequenced and published, thanks to the rapid development of genome sequencing technology [65]. Based on the high-quality genome information available, the identification of many important gene families has gradually been carried out for different plants. *DIR* proteins have been found in almost all vascular plants and often appear as a gene family [3,12]. *DIR* proteins are a kind of gene family involved in the biosynthesis of cell wall lignins and lignans, which play important roles in plant growth and development, abiotic stress tolerance, biotic stress response, and secondary metabolism regulation [16,66]. The *DIR* gene family has been successively identified in model plants, including *Arabidopsis thaliana* [16] and rice [17,18], and subsequently reported in horticultural plants such as Chinese cabbage [4], pepper [20], strawberry [21], and Cucurbitaceae [22]. As the most important solanaceous crop, eggplant is one of important vegetable crops widely cultivated across the world. However, the identification of the *DIR* gene family in eggplant has not been performed. Therefore, in this study, the *DIR* gene families were identified in eggplant based on the high-quality eggplant genome V4.1 [38]. In addition, the expression profiles of eggplant *DIR* family genes in different organs and stress responses were conducted by reanalyzing the eggplant transcriptome sequencing data, which will provide theoretical reference for further research on molecular functions of *DIR* genes and screen out the favorable *DIR* gene for eggplant resistance molecular breeding.

The number of *DIR* family genes varied greatly in different plants. In this study, a total of 24 members of the *DIR* gene family in eggplant were identified. Previous studies identified 25, 61, 35, 35, 40, 29, 24, 33, 22, 22, and 17 *DIR* family genes in *Arabidopsis* [16], rice [17,18], spruce [3,12], pear [5], *Populus* [19], Chinese cabbage [4], pepper [20], strawberry [21], watermelon, melon, and cucumber [22], respectively. The numbers of *DIR* genes in most plants that have finished the identification of their *DIR* gene family were more than the number of *DIR* genes in eggplant. The member of the *DIR* gene family in pepper was same with that of *DIR* genes in eggplant, which is probably because both pepper and eggplant belong to the crops of the Solanaceae family. The analysis of physicochemical characteristics showed that most of the *DIR* proteins in eggplant are stable proteins, which is similar to the physicochemical characteristics of *DIR* proteins in Chinese cabbage [4], pepper [20], strawberry [21], watermelon, melon, and cucumber [22]. Phylogenetic tree analysis of 24 eggplant *DIR* family genes were divided into three subgroups: *DIR-a*, *DIR-b/d*, and *DIR-e*, which was similar to the clustering results of phylogenetic analysis of *DIR* family genes in *Arabidopsis* [16], *Populus* [19], and pepper [20]. There were significant differences in gene structure among the *DIR* genes within three different subgroups, and the gene structures and conserved motifs of *DIR* genes in the same subgroup were similar. Most of eggplant *DIR* family genes (71%) only contained one exon and no introns, which were same with the exon/intron structures of *DIR* genes in other plants such as rice [17], pear [5], *Populus* [19], pepper [20], strawberry [21], watermelon, melon, and cucumber [22]. The analysis of gene duplication events of eggplant *DIR* family genes showed five pairs of tandemly duplicated genes and four pairs of segmentally duplicated genes, which revealed that the expansion of the *DIR* gene family in eggplant resulted from both tandem and segmental duplications. This phenomenon also existed in the *DIR* gene families of rice [17], pear [5], pepper [20] and strawberry [21], as well as some other gene families in other plants [67,68]. The synteny analysis of the *DIR* family genes in *Arabidopsis*, rice, and eggplant found that 15 eggplant *DIR* genes were collinear with *Arabidopsis DIR* genes, while 16 eggplant *DIR* genes were collinear with rice *DIR* genes, indicating that the eggplant *DIR* family genes have similar homology with *DIR* genes in both *Arabidopsis* and rice.

In recent years, many transcriptome sequencing projects in eggplant have been conducted with the rapid development of high-throughput sequencing technology, forming the eggplant transcriptome sequencing big data. Therefore, making good use of these transcriptome sequencing data could not only reduce the cost of research, but also enable in-depth mining of these data. Moreover, the biological functions of different gene families in eggplant could be studied using the eggplant transcriptome sequencing big data under different treatments. In this study, with the published eggplant transcriptome sequencing big data, the organ-specific expression patterns and the stress-responsive gene expression patterns of 24 eggplant *DIR* family genes were analyzed. The results showed that only *SmDIR3* was highly expressed in all 19 of the organs of eggplant. Some *SmDIR* genes, including *SmDIR7*, *SmDIR8*, *SmDIR11*, *SmDIR14*, *SmDIR18*, *SmDIR19*, *SmDIR20*, and *SmDIR23*, were not or were lowly expressed in the eggplant organs. While the other eggplant *DIR* family genes showed an organ-specific expression pattern. Previous studies have shown that *DIR* genes can contribute to the lignification of plant organs, suggesting that the *DIR* genes are essential for healthy plant growth. In our study, most of the *SmDIR* genes were specifically expressed in the radicle and root, indicating that these *SmDIR* genes play an important role in the development of radicle and root [69]. The organ-specific expression pattern of these *SmDIR* genes in different organs also cooperatively regulated the plant growth and development of eggplant.

In terms of the expression profiles of *SmDIR* genes in response to biotic and abiotic stress, 19 of 24 *SmDIR* genes were differentially expressed in response to stresses, which indicated that most of *SmDIR* genes were involved in response to stresses. Among them, five *SmDIR* genes, including *SmDIR3*, *SmDIR5*, *SmDIR6*, *SmDIR12*, and *SmDIR22*, were differentially expressed under multiple types of abiotic and biotic stresses, indicating that these five *SmDIR* genes actively participate in stress responses. Especially notable, the *SmDIR22* gene was differentially expressed under three types of abiotic stresses and two types of biotic stresses. In addition, it was also found that the expression patterns of eggplant *DIR* family genes under biotic and abiotic stresses were different. The number of differentially expressed *DIR* family genes in response to abiotic stresses was less than the number of differentially expressed *DIR* family genes in response to biotic stresses. The number of up-regulated *DIR* genes in response to biotic stresses was more than that in response to abiotic stresses. These results support the hypothesis that *DIR* genes are associated with plant-pathogen/insect infection. After the infections by pathogen and insect, more *DIR* family genes would be significantly up-regulated to accumulate more lignin, thus preventing infections by pathogen and insect. Many studies have demonstrated the importance of *DIR* genes in plant-pathogen interactions [70,71]. In addition, some eggplant *DIR* genes also play an important role in abiotic stresses, especially under high temperature, low temperature, and drought, which is consistent with the types of *cis*-elements in the promoter of *DIR* genes. Whereas, most of the eggplant *DIR* genes were down-regulated in response to stresses, which is also worthy of further investigation.

5. Conclusions

In this study, a total of 24 *DIR* family genes were identified in eggplant, which were divided into three subgroups of *DIR-a*, *DIR-b/d*, and *DIR-e*. The gene structure and protein conserved motifs in the same subgroup were highly conservative, and the distributions of exon/intron and motifs differed among different subgroups. The organ-specific expression patterns and the stress-responsive gene expression patterns of eggplant *DIR* family genes were studied by re-analyzing the eggplant transcriptome sequencing data with eggplant genome V4.1. The results showed that the expression patterns of eggplant *DIR* family genes in different organs and stress responses were different, synergistically regulating the plant growth and development of eggplant. Among the eggplant *DIR* family genes, the *SmDIR22* gene was differentially expressed under three types of abiotic stresses and two types of biotic stresses, which indicated that the *SmDIR22* gene actively participated in stress responses. The results of the current study will provide a theoretical reference for further study of the biological functions of eggplant *DIR* family genes, and highlight *SmDIR22* as the favorable gene for molecular breeding of eggplant resistance.

Author Contributions: L.J. conceived the research and designed the experiments. K.Z. performed research, analyzed the data and wrote the manuscript. W.X., S.S. and D.Y. participated in downloading transcriptome sequencing data and helped with the bioinformatics analysis. F.Z., H.J. and C.Y. analyzed and interpreted the data. All authors have read and agreed to the published version of the manuscript.

Funding: This work was supported by the National Natural Science Foundation of China (32002061), the Second Level Youth Development Fund from Anhui Academy of Agricultural Sciences (QNYC-202121), China Agriculture Research System of MOF and MARA (CSRS-23-G40), the College Students' Innovative Entrepreneurial Training Plan Program (S202110879222), the Talent Foundation of Anhui Science and Technology University (NXYJ202103), and Anhui Province Vegetable industry Technology System.

Institutional Review Board Statement: Not applicable.

Informed Consent Statement: Not applicable.

Data Availability Statement: Data used in this study are presented in the article.

Acknowledgments: We would like to Thank Martin Kagiki Njogu of Chuka University for their help in the improvement of this manuscript.

Conflicts of Interest: The authors declare no conflict of interest.

References

1. Mittler, R. Abiotic stress, the field environment and stress combination. *Trends Plant Sci.* **2006**, *11*, 15–19. [CrossRef] [PubMed]
2. Tong, H.; Madison, I.; Long, T.A.; Williams, C.M. Computational solutions for modeling and controlling plant response to abiotic stresses: A review with focus on iron deficiency. *Curr. Opin. Plant Biol.* **2020**, *57*, 8–15. [CrossRef] [PubMed]
3. Ralph, S.; Park, J.-Y.; Bohlmann, J.; Mansfield, S.D. Dirigent proteins in conifer defense: Gene discovery, phylogeny, and differential wound-and insect-induced expression of a family of DIR and DIR-like genes in spruce (*Picea* spp.). *Plant Mol. Biol.* **2006**, *60*, 21–40. [CrossRef] [PubMed]
4. Arasan, S.K.T.; Park, J.-I.; Ahmed, N.U.; Jung, H.-J.; Hur, Y.; Kang, K.-K.; Lim, Y.-P.; Nou, I.-S. Characterization and expression analysis of dirigent family genes related to stresses in *Brassica*. *Plant Physiol. Biochem.* **2013**, *67*, 144–153. [CrossRef]
5. Cheng, X.; Su, X.; Muhammad, A.; Li, M.; Zhang, J.; Sun, Y.; Li, G.; Jin, Q.; Cai, Y.; Lin, Y. Molecular characterization, evolution, and expression profiling of the *dirigent* (DIR) family genes in Chinese white pear (*Pyrus bretschneideri*). *Front. Genet.* **2018**, *9*, 136. [CrossRef]
6. Bonello, P.; Storer, A.J.; Gordon, T.R.; Wood, D.L.; Heller, W. Systemic effects of *Heterobasidion annosum* on ferulic acid glucoside and lignin of presymptomatic ponderosa pine phloem, and potential effects on bark-beetle-associated fungi. *J. Chem. Ecol.* **2003**, *29*, 1167–1182. [CrossRef]
7. Miedes, E.; Vanholme, R.; Boerjan, W.; Molina, A. The role of the secondary cell wall in plant resistance to pathogens. *Front. Plant Sci.* **2014**, *5*, 358. [CrossRef]
8. Davin, L.B.; Jourdes, M.; Patten, A.M.; Kim, K.-W.; Vassao, D.G.; Lewis, N.G. Dissection of lignin macromolecular configuration and assembly: Comparison to related biochemical processes in allyl/propenyl phenol and lignan biosynthesis. *Nat. Prod. Rep.* **2008**, *25*, 1015–1090. [CrossRef]
9. Davin, L.B.; Lewis, N.G. Lignin primary structures and dirigent sites. *Curr. Opin. Biotech* **2005**, *16*, 407–415. [CrossRef]
10. MacRae, W.D.; Towers, G.N. Biological activities of lignans. *Phytochemistry* **1984**, *23*, 1207–1220. [CrossRef]
11. Harmatha, J.; Dinan, L. Biological activities of lignans and stilbenoids associated with plant-insect chemical interactions. *Phytochem. Rev.* **2003**, *2*, 321–330. [CrossRef]
12. Ralph, S.G.; Jancsik, S.; Bohlmann, J. Dirigent proteins in conifer defense II: Extended gene discovery, phylogeny, and constitutive and stress-induced gene expression in spruce (*Picea* spp.). *Phytochemistry* **2007**, *68*, 1975–1991. [CrossRef] [PubMed]
13. Wu, R.; Wang, L.; Wang, Z.; Shang, H.; Liu, X.; Zhu, Y.; Qi, D.; Deng, X. Cloning and expression analysis of a dirigent protein gene from the resurrection plant *Boea hygrometrica*. *Prog. Nat. Sci.* **2009**, *19*, 347–352. [CrossRef]
14. Li, Q.; Chen, J.; Xiao, Y.; Di, P.; Zhang, L.; Chen, W. The dirigent multigene family in *Isatis indigotica*: Gene discovery and differential transcript abundance. *BMC Genom.* **2014**, *15*, 388. [CrossRef] [PubMed]
15. Davin, L.B.; Wang, H.-B.; Crowell, A.L.; Bedgar, D.L.; Martin, D.M.; Sarkanen, S.; Lewis, N.G. Stereoselective bimolecular phenoxy radical coupling by an auxiliary (dirigent) protein without an active center. *Science* **1997**, *275*, 362–367. [CrossRef]
16. Paniagua, C.; Bilkova, A.; Jackson, P.; Dabravolski, S.; Riber, W.; Didi, V.; Houser, J.; Gigli-Bisceglia, N.; Wimmerova, M.; Budínská, E. Dirigent proteins in plants: Modulating cell wall metabolism during abiotic and biotic stress exposure. *J. Exp. Bot.* **2017**, *68*, 3287–3301. [CrossRef]
17. Liao, Y.; Liu, S.; Jiang, Y.; Hu, C.; Zhang, X.; Cao, X.; Xu, Z.; Gao, X.; Li, L.; Zhu, J. Genome-wide analysis and environmental response profiling of dirigent family genes in rice (*Oryza sativa*). *Genes Genom.* **2017**, *39*, 47–62. [CrossRef]
18. Singh, D.K.; Mehra, S.; Chatterjee, S.; Purty, R.S. In silico identification and validation of miRNA and their DIR specific targets in *Oryza sativa* Indica under abiotic stress. *Non-Coding RNA Res.* **2020**, *5*, 167–177. [CrossRef]
19. Li, L.; Sun, W.; Zhou, P.; Wei, H.; Wang, P.; Li, H.; Rehman, S.; Li, D.; Zhuge, Q. Genome-wide characterization of dirigent proteins in *Populus*: Gene expression variation and expression pattern in response to *Marssonina brunnea* and phytohormones. *Forests* **2021**, *12*, 507. [CrossRef]
20. Khan, A.; Li, R.-J.; Sun, J.-T.; Ma, F.; Zhang, H.-X.; Jin, J.-H.; Ali, M.; Wang, J.-E.; Gong, Z.-H. Genome-wide analysis of dirigent gene family in pepper (*Capsicum annuum* L.) and characterization of *CaDIR7* in biotic and abiotic stresses. *Sci. Rep.* **2018**, *8*, 5500. [CrossRef]
21. Shi, Y.; Shen, Y.; Ahmad, B.; Yao, L.; He, T.; Fan, J.; Liu, Y.; Chen, Q.; Wen, Z. Genome-wide identification and expression analysis of dirigent gene family in strawberry (*Fragaria vesca*) and functional characterization of *FvDIR13*. *Sci. Hortic.* **2022**, *297*, 110913. [CrossRef]
22. Yadav, V.; Wang, Z.; Yang, X.; Wei, C.; Changqing, X.; Zhang, X. Comparative analysis, characterization and evolutionary study of dirigent gene family in cucurbitaceae and expression of novel dirigent peptide against powdery mildew stress. *Genes* **2021**, *12*, 326. [CrossRef] [PubMed]
23. Jin-Long, G.; Li-Ping, X.; Jing-Ping, F.; Ya-Chun, S.; Hua-Ying, F.; You-Xiong, Q.; Jing-Sheng, X. A novel dirigent protein gene with highly stem-specific expression from sugarcane, response to drought, salt and oxidative stresses. *Plant Cell Rep.* **2012**, *31*, 1801–1812. [CrossRef] [PubMed]
24. Andrade, L.M.; Peixoto-Junior, R.F.; Ribeiro, R.V.; Nóbile, P.M.; Brito, M.S.; Marchiori, P.E.R.; Carlin, S.D.; Martins, A.P.B.; Goldman, M.H.S.; Llerena, J.P.P. Biomass accumulation and cell wall structure of rice plants overexpressing a dirigent-jacalin of sugarcane (*ShDJ*) under varying conditions of water availability. *Front. Plant Sci.* **2019**, *10*, 65. [CrossRef] [PubMed]
25. Liu, C.; Qin, Z.; Zhou, X.; Xin, M.; Wang, C.; Liu, D.; Li, S. Expression and functional analysis of the Propamocarb-related gene *CsDIR16* in cucumbers. *BMC Plant Biol.* **2018**, *18*, 16. [CrossRef] [PubMed]

26. Wang, Y.; Cao, Y.; Liang, X.; Zhuang, J.; Wang, X.; Qin, F.; Jiang, C. A dirigent family protein confers variation of Casparian strip thickness and salt tolerance in maize. *Nat. Commun.* **2022**, *13*, 2222. [CrossRef]
27. Burlat, V.; Kwon, M.; Davin, L.B.; Lewis, N.G. Dirigent proteins and dirigent sites in lignifying tissues. *Phytochemistry* **2001**, *57*, 883–897. [CrossRef]
28. Li, N.; Zhao, M.; Liu, T.; Dong, L.; Cheng, Q.; Wu, J.; Wang, L.; Chen, X.; Zhang, C.; Lu, W. A novel soybean dirigent gene *GmDIR22* contributes to promotion of lignan biosynthesis and enhances resistance to *Phytophthora sojae*. *Front. Plant Sci.* **2017**, *8*, 1185. [CrossRef]
29. Seneviratne, H.K.; Dalisay, D.S.; Kim, K.-W.; Moinuddin, S.G.; Yang, H.; Hartshorn, C.M.; Davin, L.B.; Lewis, N.G. Non-host disease resistance response in pea (*Pisum sativum*) pods: Biochemical function of DRR206 and phytoalexin pathway localization. *Phytochemistry* **2015**, *113*, 140–148. [CrossRef] [PubMed]
30. Reboledo, G.; Del Campo, R.; Alvarez, A.; Montesano, M.; Mara, H.; Ponce de León, I. *Physcomitrella patens* activates defense responses against the pathogen *Colletotrichum gloeosporioides*. *Int. J. Mol. Sci.* **2015**, *16*, 22280–22298. [CrossRef]
31. Borges, A.F.; Ferreira, R.B.; Monteiro, S. Transcriptomic changes following the compatible interaction *Vitis vinifera-Erysiphe necator*. Paving the way towards an enantioselective role in plant defence modulation. *Plant Physiol. Biochem.* **2013**, *68*, 71–80. [CrossRef] [PubMed]
32. Rotino, G.L.; Sala, T.; Toppino, L. Eggplant. In *Alien Gene Transfer in Crop Plants, Volume 2*; Springer: New York, NY, USA, 2014; pp. 381–409.
33. Chapman, M.A. Introduction: The importance of eggplant. In *The Eggplant Genome*; Springer: Cham, Switzerland, 2019; pp. 1–10.
34. Hirakawa, H.; Shirasawa, K.; Miyatake, K.; Nunome, T.; Negoro, S.; Ohyama, A.; Yamaguchi, H.; Sato, S.; Isobe, S.; Tabata, S. Draft genome sequence of eggplant (*Solanum melongena* L.): The representative *solanum* species indigenous to the old world. *DNA Res.* **2014**, *21*, 649–660. [CrossRef] [PubMed]
35. Barchi, L.; Pietrella, M.; Venturini, L.; Minio, A.; Toppino, L.; Acquadro, A.; Andolfo, G.; Aprea, G.; Avanzato, C.; Bassolino, L. A chromosome-anchored eggplant genome sequence reveals key events in Solanaceae evolution. *Sci. Rep.* **2019**, *9*, 11769. [CrossRef] [PubMed]
36. Wei, Q.; Wang, J.; Wang, W.; Hu, T.; Hu, H.; Bao, C. A high-quality chromosome-level genome assembly reveals genetics for important traits in eggplant. *Hortic. Res.* **2020**, *7*, 153. [CrossRef]
37. Li, D.; Qian, J.; Li, W.; Yu, N.; Gan, G.; Jiang, Y.; Li, W.; Liang, X.; Chen, R.; Mo, Y. A high-quality genome assembly of the eggplant provides insights into the molecular basis of disease resistance and chlorogenic acid synthesis. *Mol. Ecol. Resour.* **2021**, *21*, 1274–1286. [CrossRef]
38. Barchi, L.; Rabanus-Wallace, M.T.; Prohens, J.; Toppino, L.; Padmarasu, S.; Portis, E.; Rotino, G.L.; Stein, N.; Lanteri, S.; Giuliano, G. Improved genome assembly and pan-genome provide key insights into eggplant domestication and breeding. *Plant J.* **2021**, *107*, 579–596. [CrossRef]
39. Wang, S.; Chen, Z.; Ji, T.; Di, Q.; Li, L.; Wang, X.; Wei, M.; Shi, Q.; Li, Y.; Gong, B. Genome-wide identification and characterization of the R2R3MYB transcription factor superfamily in eggplant (*Solanum melongena* L.). *Agri Gene* **2016**, *2*, 38–52. [CrossRef]
40. Yang, Y.; Liu, J.; Zhou, X.; Liu, S.; Zhuang, Y. Identification of WRKY gene family and characterization of cold stress-responsive WRKY genes in eggplant. *PeerJ* **2020**, *8*, e8777. [CrossRef]
41. Wan, F.-x.; Gao, J.; Wang, G.-l.; Niu, Y.; Wang, L.-z.; Zhang, X.-g.; Wang, Y.-q.; Pan, Y. Genome-wide identification of NAC transcription factor family and expression analysis of ATAF subfamily members under abiotic stress in eggplant. *Sci. Hortic.* **2021**, *289*, 110424. [CrossRef]
42. Chen, J.; Wang, S.; Wu, F.; Wei, M.; Li, J.; Yang, F. Genome-wide identification and functional characterization of auxin response factor (ARF) genes in eggplant. *Int. J. Mol. Sci.* **2022**, *23*, 6219. [CrossRef]
43. Finn, R.D.; Clements, J.; Eddy, S.R. HMMER web server: Interactive sequence similarity searching. *Nucleic Acids Res.* **2011**, *39*, W29–W37. [CrossRef] [PubMed]
44. Chen, C.; Chen, H.; Zhang, Y.; Thomas, H.R.; Frank, M.H.; He, Y.; Xia, R. TBtools: An integrative toolkit developed for interactive analyses of big biological data. *Mol. Plant* **2020**, *13*, 1194–1202. [CrossRef] [PubMed]
45. Mistry, J.; Chuguransky, S.; Williams, L.; Qureshi, M.; Salazar, G.A.; Sonnhammer, E.L.; Tosatto, S.C.; Paladin, L.; Raj, S.; Richardson, L.J. Pfam: The protein families database in 2021. *Nucleic Acids Res.* **2021**, *49*, D412–D419. [CrossRef] [PubMed]
46. Letunic, I.; Khedkar, S.; Bork, P. SMART: Recent updates, new developments and status in 2020. *Nucleic Acids Res.* **2021**, *49*, D458–D460. [CrossRef]
47. Yu, C.S.; Lin, C.J.; Hwang, J.K. Predicting subcellular localization of proteins for Gram-negative bacteria by support vector machines based on *n*-peptide compositions. *Protein Sci.* **2004**, *13*, 1402–1406. [CrossRef]
48. Bailey, T.L.; Williams, N.; Misleh, C.; Li, W.W. MEME: Discovering and analyzing DNA and protein sequence motifs. *Nucleic Acids Res.* **2006**, *34*, W369–W373. [CrossRef]
49. Lescot, M.; Déhais, P.; Thijs, G.; Marchal, K.; Moreau, Y.; Van de Peer, Y.; Rouzé, P.; Rombauts, S. PlantCARE, a database of plant *cis*-acting regulatory elements and a portal to tools for in silico analysis of promoter sequences. *Nucleic Acids Res.* **2002**, *30*, 325–327. [CrossRef]
50. Tamura, K.; Stecher, G.; Kumar, S. MEGA11: Molecular evolutionary genetics analysis version 11. *Mol. Biol. Evol.* **2021**, *38*, 3022–3027. [CrossRef]

51. Subramanian, B.; Gao, S.; Lercher, M.J.; Hu, S.; Chen, W.-H. Evolview v3: A webserver for visualization, annotation, and management of phylogenetic trees. *Nucleic Acids Res.* **2019**, *47*, W270–W275. [CrossRef]
52. Wang, Y.; Tang, H.; DeBarry, J.D.; Tan, X.; Li, J.; Wang, X.; Lee, T.-h.; Jin, H.; Marler, B.; Guo, H. MCScanX: A toolkit for detection and evolutionary analysis of gene synteny and collinearity. *Nucleic Acids Res.* **2012**, *40*, e49. [CrossRef]
53. Krzywinski, M.; Schein, J.; Birol, I.; Connors, J.; Gascoyne, R.; Horsman, D.; Jones, S.J.; Marra, M.A. Circos: An information aesthetic for comparative genomics. *Genome Res.* **2009**, *19*, 1639–1645. [CrossRef] [PubMed]
54. Bolger, A.M.; Lohse, M.; Usadel, B. Trimmomatic: A flexible trimmer for Illumina sequence data. *Bioinformatics* **2014**, *30*, 2114–2120. [CrossRef] [PubMed]
55. Dobin, A.; Davis, C.A.; Schlesinger, F.; Drenkow, J.; Zaleski, C.; Jha, S.; Batut, P.; Chaisson, M.; Gingeras, T.R. STAR: Ultrafast universal RNA-seq aligner. *Bioinformatics* **2013**, *29*, 15–21. [CrossRef] [PubMed]
56. Li, H.; Handsaker, B.; Wysoker, A.; Fennell, T.; Ruan, J.; Homer, N.; Marth, G.; Abecasis, G.; Durbin, R. The sequence alignment/map format and SAMtools. *Bioinformatics* **2009**, *25*, 2078–2079. [CrossRef]
57. Pertea, M.; Pertea, G.M.; Antonescu, C.M.; Chang, T.-C.; Mendell, J.T.; Salzberg, S.L. StringTie enables improved reconstruction of a transcriptome from RNA-seq reads. *Nat. Biotechnol.* **2015**, *33*, 290–295. [CrossRef]
58. Love, M.I.; Huber, W.; Anders, S. Moderated estimation of fold change and dispersion for RNA-seq data with DESeq2. *Genome Biol.* **2014**, *15*, 550. [CrossRef]
59. Lv, L.L.; Feng, X.F.; Li, W.; Li, K. High temperature reduces peel color in eggplant (*Solanum melongena*) as revealed by RNA-seq analysis. *Genome* **2019**, *62*, 503–512. [CrossRef]
60. Yang, Y.; Liu, J.; Zhou, X.; Liu, S.; Zhuang, Y. Transcriptomics analysis unravels the response to low temperature in sensitive and tolerant eggplants. *Sci. Hortic.* **2020**, *271*, 109468. [CrossRef]
61. Li, J.; Gao, Z.; Zhou, L.; Li, L.; Zhang, J.; Liu, Y.; Chen, H. Comparative transcriptome analysis reveals K+ transporter gene contributing to salt tolerance in eggplant. *BMC Plant Biol.* **2019**, *19*, 67. [CrossRef]
62. Peng, J.; Wang, P.; Fang, H.; Zheng, J.; Zhong, C.; Yang, Y.; Yu, W. Weighted gene co-expression analysis network-based analysis on the candidate pathways and hub genes in eggplant bacterial wilt-resistance: A plant research study. *Int. J. Mol. Sci.* **2021**, *22*, 13279. [CrossRef]
63. Yang, X.; Zhang, Y.; Cheng, Y.; Chen, X. Transcriptome analysis reveals multiple signal network contributing to the *Verticillium wilt* resistance in eggplant. *Sci. Hortic.* **2019**, *256*, 108576. [CrossRef]
64. Chen, L.-m.; Li, X.-w.; He, T.-j.; Li, P.-j.; Liu, Y.; Zhou, S.-x.; Wu, Q.-c.; Chen, T.-t.; Lu, Y.-b.; Hou, Y.-m. Comparative biochemical and transcriptome analyses in tomato and eggplant reveal their differential responses to *Tuta absoluta* infestation. *Genomics* **2021**, *113*, 2108–2121. [CrossRef] [PubMed]
65. Sun, Y.; Shang, L.; Zhu, Q.-H.; Fan, L.; Guo, L. Twenty years of plant genome sequencing: Achievements and challenges. *Trends Plant Sci.* **2021**, *27*, 391–401. [CrossRef] [PubMed]
66. Davin, L.B.; Lewis, N.G. Dirigent proteins and dirigent sites explain the mystery of specificity of radical precursor coupling in lignan and lignin biosynthesis. *Plant Physiol.* **2000**, *123*, 453–462. [CrossRef]
67. Cannon, S.B.; Mitra, A.; Baumgarten, A.; Young, N.D.; May, G. The roles of segmental and tandem gene duplication in the evolution of large gene families in *Arabidopsis thaliana*. *BMC Plant Biol.* **2004**, *4*, 10. [CrossRef]
68. Zhu, Y.; Wu, N.; Song, W.; Yin, G.; Qin, Y.; Yan, Y.; Hu, Y. Soybean (*Glycine max*) expansin gene superfamily origins: Segmental and tandem duplication events followed by divergent selection among subfamilies. *BMC Plant Biol.* **2014**, *14*, 93. [CrossRef]
69. Liu, Q.; Luo, L.; Zheng, L. Lignins: Biosynthesis and biological functions in plants. *Int. J. Mol. Sci.* **2018**, *19*, 335. [CrossRef]
70. Alvarez, A.; Montesano, M.; Schmelz, E.; Ponce de León, I. Activation of shikimate, phenylpropanoid, oxylipins, and auxin pathways in *Pectobacterium carotovorum* elicitors-treated moss. *Front. Plant Sci.* **2016**, *7*, 328. [CrossRef]
71. Weidenbach, D.; Esch, L.; Möller, C.; Hensel, G.; Kumlehn, J.; Höfle, C.; Hückelhoven, R.; Schaffrath, U. Polarized defense against fungal pathogens is mediated by the Jacalin-related lectin domain of modular *Poaceae*-specific proteins. *Mol. Plant* **2016**, *9*, 514–527. [CrossRef]

horticulturae

Article

Molecular Cloning, Characterization, and Expression Analysis of SIMILAR TO RCD-ONE (SRO) Family Genes Responding to Abiotic and Biotic Stress in Cucumber

Lingdi Xiao [1,†], Zixian Zhou [1,†], Chuxia Zhu [1], Jindong Zhao [1], Zhaoyang Hu [1], Shiqiang Liu [1,*] and Yong Zhou [1,2,*]

1 College of Bioscience and Bioengineering, Jiangxi Agricultural University, Nanchang 330045, China; xiaolingdixiao@163.com (L.X.); zixianzhou@163.com (Z.Z.); z18979935173@163.com (C.Z.); zjd13519442751@163.com (J.Z.); huzhaoyang@jxau.edu.cn (Z.H.)
2 Key Laboratory of Crop Physiology, Ecology and Genetic Breeding, Ministry of Education, Jiangxi Agricultural University, Nanchang 330045, China
* Correspondence: lsq_hn306@163.com (S.L.); yongzhou@jxau.edu.cn (Y.Z.)
† These authors contributed equally to this work.

Abstract: SIMILAR TO RCD-ONE (SRO) is a plant-specific small protein family that controls many biological processes including physiological development and stress responses. The *SRO* gene family has been studied in several plant species, but no detailed characterization and expression profiles of this important gene family were performed in cucumber. In this study, we characterize the *SRO* genes in cucumber, and determined their transcript levels in various tissues and under exposure to diverse biotic and abiotic stressors. Four *SRO* genes (named as *CsSRO1–CsSRO4*) were identified and isolated, which were distributed on three different chromosomes. Gene duplication analysis showed that only one pair of segmental duplication event was identified, but no tandem duplication events were detected. All CsSROs consist of the PARP domain and a C-terminal RST domain, while the N-terminal WWE domain was only present in CsSRO2 and CsSRO4. SROs from 15 plant species are divided into two groups (I and II), and group I can be further divided into four subgroups (Ia to Id) according to the phylogenetic tree. The conserved motif and gene structure analyses showed that SROs within the same branch of the phylogenetic tree have analogous conserved motifs configuration and gene structures. However, *SRO* genes possessed variable numbers of introns in different subgroups, which may affect the evolution of new family members. RNA-Seq data and qRT-PCR results showed that the four *CsSRO* genes have distinct expression pattern in various tissues and under diverse stresses, suggesting their multiple functions in plant growth and stress responses. The findings provide a basis for further research aiming at functional characterization of the regulatory mechanism to reveal the roles of *CsSRO* genes in developmental and stress-related processes of cucumber.

Keywords: cucumber; *SRO* gene; stress response; gene expression; gene family

1. Introduction

Plants have produced a series of regulatory networks through the transcriptional control of numerous stress-responsive genes to protect themselves from adverse environmental effects [1,2]. Transcription factors (TFs) are a class of proteins that can bind to the *cis*-regulatory elements in promoters of target genes and thus to activate or inhibit their expression, thereby regulating plant development and the response to various stresses [3,4]. SIMILAR TO RCD-ONE (SRO) is a unique type of small TF proteins in plants, which can act as key regulators involved in regulation of many biological pathways. Reports have found that SRO proteins consist of a conserved poly (ADP-ribose) polymerase catalytic center (PARP domain, PF00644) and a C-terminal RCD-SRO-TAF4 domain (RST domain, PF12174) that interacts with other TFs. Some SROs also contain a WWE domain (PF02825) at the N-terminus, which takes part in ADP ribosylation (ADPRylation) and ubiquitination of proteins [5–7].

The first identified *SRO* family member, *RADICAL-INDUCED CELL DEATH1* (*RCD1*) in *Arabidopsis*, was able to complement the yeast *yap1(-)* mutant which has defects in oxidative stress sensitivity, suggesting that it might participate in oxidative stress [8]. AtRCD1 also has a role in salt tolerance through physically interacting with the plasma membrane localization Na^+/H^+ antiporter SALT OVERLY SENSITIVE 1 (SOS1) [9]. The *rcd1* mutants exhibited pleiotropic phenotypes such as the aberrant leaf, root and rosette development, early flowering, and altered resistance to diverse abiotic stresses and hormones response [10–15]. Subsequently, five homologs (named as AtSRO1–5) of AtRCD1 were also identified in *Arabidopsis*. AtSRO1 is the closest homolog of AtRCD1, both of them contained WWE domain at the N-terminus and possessed similar expression patterns, but they play unequally redundant and independent roles in certain developmental processes and the response to various stresses [14,16–18]. For example, *sro1-1* mutants displayed only unconspicuous morphological phenotypes, while *rcd1-3 sro1-1* double mutants showed even more extreme phenotypes than *rcd1-3* [13,14]. Additionally, *rcd1* and *sro1* mutants also showed different responses to oxidative stress [13], but they respond similarly to $HgCl_2$ and H_2O_2 stress [19]. Another three *SRO* family members, *AtSRO2*, *AtSRO3*, and *AtSRO5*, displayed changed transcriptional levels under strong light, salt, $HgCl_2$, H_2O_2, and ozone stress conditions. In addition, overexpression of *AtSRO5* conferred salt tolerance of transgenic plants [18–20].

Besides *Arabidopsis*, *SRO* genes have also been extensively identified and functionally studied in other plant species. In rice, *OsSRO1c* functions as a core regulator to mediate abiotic stress responses via direct interaction with various TFs, such as SNAC1 (stress-responsive NAC 1) and DST (drought and salt tolerance) [21,22]. As an allele of *OsSRO1c* from the *indica* rice cultivar (*Oryza sativa* ssp. *indica*), *BROWNING OF CALLUS1* (*BOC1*) can improve genetic transformation efficiency probably through inhibiting programmed cell death caused by oxidative stress, and thus reducing callus browning [23]. Another SRO member in rice, OsSRO1a, negatively regulates the rice-*Xanthomonas oryzae* interaction via the OsMYC2-mediated jasmonic acid (JA) signaling pathway [24]. A bread wheat (*Triticum aestivum*) *SRO* gene *TaSRO1* showed promoted plant growth and multiple abiotic stress tolerance by controlling redox homoeostasis in transgenic wheat and *Arabidopsis* lines [25]. Overexpression of maize *ZmSRO1b* in *Arabidopsis* also elevated the tolerance of plants to salt, cadmium, and oxidative stress [26]. In contrast, *ZmSRO1e*-overexpressing maize and *Arabidopsis* plants exhibited increased sensitivity to oxidative and salt stress with reduced anthocyanin accumulation through disrupting the MYB–bHLH–WD40 (MBW) complex [27]. These findings demonstrated that *SRO* genes play vital roles in plant development, and response to multiple biotic and abiotic stresses.

Cucumber is an important vegetable crop with great economic value in the world, but it always encounters a variety of environmental stimulus (such as drought, salt, and pathogens) during its growth cycle [4,28]. In recent years, many stress-responsive TF gene families have been identified in cucumber, such as GARS [29,30], bHLH [31], WRKY [32], VQ [33], and TIFY [34], and some members of them were proven to function in stress responses [35,36]. However, no report has been carried out on the *SRO* family genes in cucumber until now. To elucidate whether the cucumber *SRO* family genes participate in stress response, we cloned and characterized four cucumber *SRO* genes, and the phylogenetic relationship, conserved motif, gene structure, and promoter region were analyzed by some routine bioinformatics tools. Additionally, the transcriptional levels of cucumber *SRO* genes under salt and drought stress conditions, as well as under the infection of *Pseudoperonospora cubensis* and *Meloidogyne incognita*, were also determined by qRT-PCR and RNA-seq analyses. Our findings provide a solid base for the classification and expression of the *SRO* family genes and help to facilitate further functional characterization of these genes in cucumber.

2. Materials and Methods

2.1. Plant Materials, Growth Conditions and Stress Treatments

The Chinese Long cucumber inbred line 9930 plants were employed for stress treatments in this work. The seeds were soaked in water at 55 °C for 15 min, then placed on damp filter paper overnight in an artificial climate incubator at 28 °C to promote germination. After the cotyledons were fully opened, the seedlings were moved to Hoagland's nutrition for cultivation. Cucumber seedlings were soaked in Hoagland nutrient solution for hydroponics with a 16:8 h light: dark photoperiod. We transferred the two-leaf stage seedlings to a nutrition solution containing NaCl (200 mM) and PEG6000 (10%), for salt stress and drought stress, respectively, as described previously [4]. The leaf tissues were harvested for each stress condition with three biological replicates at 0, 3, 6, and 12 h after treatment. Subsequently, these samples were stored in a −80 °C freezer for RNA extraction until further experiments.

2.2. Identification and Cloning of CsSRO Genes of Cucumber

The Hidden Markov Model (HMM) profiles of the RST domain (PF12174) and PARP domain (PF00644) were downloaded from the Pfam database (http://pfam.xfam.org/, accessed on 19 June 2022) [37], and these HMM profiles were employed to search against the cucumber protein database downloaded from Cucurbit Genomics Database (CuGenDB) (http://cucurbitgenomics.org/, accessed on 19 June 2022) using HMMER software. We also searched the cucumber *SRO* family members using the BLASTP with the published *Arabidopsis* and rice SRO protein sequences as query sequences. After removing the redundant sequences obtained from the above methods and checking through Pfam and InterProScan (http://www.ebi.ac.uk/Tools/InterProScan/, accessed on 19 June 2022), we get four candidate SRO protein sequences, and their coding sequences (CDS) and corresponding genomic DNA (gDNA) sequences were also obtained from CuGenDB.

Total RNA was isolated from the leaf samples of normally growing and stress-treated seedlings using RNA-easy™ Isolation Reagent Vazyme cat (Vazyme Biology Co., Ltd., Nanjing, China), and the first strand cDNAs were synthesized using the HiScript®II Q RT SuperMix for qPCR (+gDNA wipe) (Vazyme Biology Co., Ltd., Nanjing, China) on the basis of to the manufacturer's instructions. The cDNA samples were diluted 10 times, and used as the template for PCR amplification. The PCR amplification was carried out at: denaturation at 95 °C for 3 min; 95 °C for 1 min, 60 °C for 15 s, and 72 °C for 3 min (35 cycles); followed by 72 °C for 5 min. The PCR products were detected by 1.0% agarose gel electrophoresis and subsequently sequencing. Primer sequences used for PCR amplification of each *CsSRO* gene are listed in Table S1.

2.3. In Silico Sequence Analyses of Cucumber SRO Family Members

The physicochemical characteristics of cucumber SRO proteins, including molecular weight (MW), isoelectric point (pI), and grand average of hydropathicity (GRAVY), were calculated using the ProtParam tool on the ExPASy server (http://web.expasy.org/protparam, accessed on 19 June 2022) [38]. The subcellular location of cucumber SRO proteins was predicted using CELLO v.2.5 (http://cello.life.nctu.edu.tw/, accessed on 19 June 2022) [39]. The conserved domains including PARP, RST, and WWE were examined through Pfam and InterProScan. The physical locations of *CsSRO* genes were illustrated with the MapInspect software according to the initial position information provided in CuGenDB. Moreover, these genes involved in the tandem or segmental duplications were analyzed with MCScanX on the basis of the criterion in our previous report [4]. The 2000 bp promoter sequences located upstream of the initiation codon (ATG) of the cucumber *SRO* genes were retrieved from CuGenDB, and various *cis*-regulatory elements related to stresses and hormones were determined using PlantCARE (http://bioinformatics.psb.ugent.be/webtools/plantcare/html/, accessed on 19 June 2022) [40].

2.4. Phylogenetic, Conserved Motif and Gene Structure Analyses of SRO Family Members of Cucumber and Other Plant Species

The full-length protein sequence alignments of all *SRO* family members from cucumber and different plant species were performed using the MAFFT tool under default settings, and then MEGA version 7.0 was used to construct phylogenetic trees by the Neighbor-Joining (NJ) methods and the bootstrap analysis conducted with 1000 replications [41]. MEME (http://meme-suite.org/tools/meme, accessed on 19 June 2022) was employed to examine the conserved motifs of SRO proteins of cucumber and other plant species with the default settings except the number of diverse motifs was chosen as 10 [42]. The gene structures of the *SRO* genes of cucumber, *Arabidopsis*, maize, rice, and grape were determined with the GSDS 2.0 tool (Gene Structure Display Server, http://gsds.gao-lab.org/, accessed on 19 June 2022) according to their CDS and the correspondent full-length gDNA sequences [43].

2.5. qRT-PCR Analysis of CsSRO Genes in Response to Abiotic Stress

The quantitative RT-PCR was conducted using the Roche Lightcyler 480II PCR System based on the manufacturer's instructions of the ChamQ Universal SYBR qPCR Master Mix Kit (Vazyme Biology Co., Ltd., Nanjing, China). The total qRT-PCR mixture was 20 µL including 10 µL 2 × ChamQ Universal SYBR qPCR Master Mix, 0.4 µL forward primer, 0.4 µL reverse primer, 7.2 µL ddH_2O, and 2 µL template cDNA. The reaction conditions were as follows: 95 °C for 30 s, followed by 40 cycles of 95 °C for 10 s, and 60 °C for 30 s. All reactions were performed in triplicate, and *CsAct3* was used as the reference gene to determine the relative expression with the $2^{-\Delta\Delta Ct}$ method [4]. qRT-PCR-specific primers used for expression analysis of each *CsSRO* gene were listed in Table S1.

2.6. Expression Analysis of CsSRO Genes Based on RNA-seq

The expression data of *CsSRO* genes in various tissues, including root, leaf, stem, female and male flowers, ovaries, tendril, and the base of tendril, were retrieved from the RNA-seq in NCBI under the accession number of PRJNA80169. The expression data of *CsSRO* genes in different biotic stresses, including the infection of *P. cubensis* (downy mildew treatment) and *M. incognita* (root-knot nematode treatment), were available in the NCBI under the accession numbers of PRJNA388584 and PRJNA419665, respectively. Based on all expression values, TBtools was used to generate expression heatmaps using the $\log_2$ transformed transcripts per million reads (TPM) values, based on the methods in our previous studies [4,44].

3. Results

3.1. Cloning and Identification of SRO Gene Sequences of Cucumber

A total of four *SRO* genes were identified in cucumber, and they were numbered from *CsSRO1* to *CsSRO4*, on the basis of their chromosomal locations (Table 1). To verify the accuracy of sequences acquired from the cucumber genome, we designed the specific primers based on the CDS sequences of *CsSRO* genes. The 1167 bp, 1398 bp, 1806 bp, and 1863 bp CDSs of the *CsSRO* genes were amplified, by using cDNA from cucumber leaf tissues as a template (Figure S1). Sequencing of PCR products confirmed the coding sequences of *CsSRO* genes, suggesting the reliability of the sequence information of the above identification (Table 1). The physicochemical characteristics including sequence length, MW, pI, and GRAVY values of CsSROs are provided in Table 1.

Table 1. Identification and characterization of *SRO* family genes in cucumber.

Gene	Locus ID	Chromosomal Position	Protein					gDNA (bp)	CDS (bp)
			Length (aa)	MW (Da)	pI	GRAVY	Subcellular Prediction		
CsSRO1	Csa1G046060	Chr1: 5287727-5290312	388	43,102.56	7.75	0.511	Nucleus	2586	1167
CsSRO2	Csa2G372800	Chr2: 18564803-18569546	465	51,712.31	9.33	0.324	Mitochondria/Nucleus	4744	1398
CsSRO3	Csa3G008310	Chr3: 918184-923735	601	67,660.98	5.79	0.373	Nucleus	5552	1806
CsSRO4	Csa3G901000	Chr3: 38763905-38773671	620	70,266.98	7.55	0.502	Nucleus	9767	1863

The basic characteristic analysis showed that the gDNA and CDS lengths of four *CsSRO* genes varied in size, with gDNA lengths varied from 2586 bp (*CsSRO1*) to 9767 bp (*CsSRO4*) and CDS lengths ranged from 1167 bp (*CsSRO1*) to 1863 bp (*CsSRO4*) (Table 1). Their gene products had molecular weight (MW) and isoelectric points (pI) with MW ranging from 43,102.56 Da (CsSRO1) to 70,266.98 Da (CsSRO4) and pI ranging from 5.79 (CsSRO3) to 9.33 (CsSRO2). The GRAVY values of CsSROs ranged from −0.511 (CsSRO1) to −0.324 (CsSRO2), which were less than zero, suggesting that the CsSRO proteins were hydrophilic proteins. The subcellular localization prediction results showed that CsSRO1, CsSRO3, and CsSRO4 were predicted to be localized in the nucleus, while CsSRO2 might be located in mitochondria and nucleus (Table 1).

3.2. Phylogenetic Analysis of SRO Family Genes from Cucumber and Different Plant Species

To investigate the phylogenetic relationship and divergence of the *SRO* family in cucumber, a phylogenetic tree was created on the basis of multiple sequence alignment of protein sequences of *SRO* gene family from cucumber and other 14 different plant species. As shown in the phylogenetic tree, these SRO proteins were irregularly divided into two groups, namely group I and group II, and group I can be further divided into four subgroups (Ia to Id) (Figure 1). Consistent with previous reports [45,46], the SRO proteins of these species were clustered in dicot- or monocot-specific patterns. For example, group II had only dicotyledonous SRO proteins, while group I contained both monocotyledonous and dicotyledonous SRO proteins. For CsSRO proteins, CsSRO3 and CsSRO4 were clustered in group Ic, while CsSRO3 and CsSRO4 were clustered in group Id and group II, respectively (Figure 1).

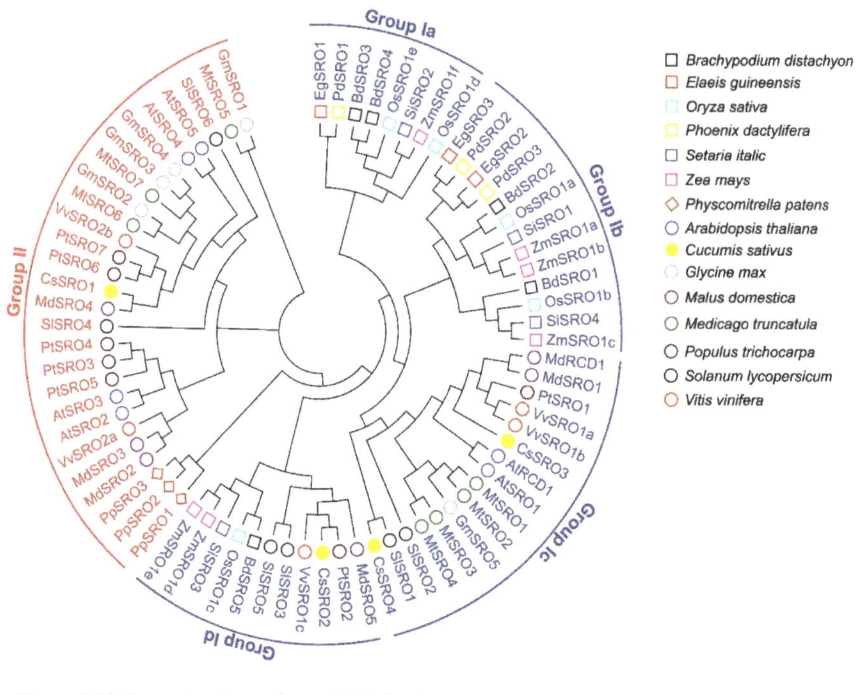

Figure 1. Phylogenetic relationships of *SRO* family genes from cucumber and 14 other plant species. The phylogenetic tree was created with MEGA 7.0 using the Neighbor-Joining (NJ) method with bootstrap analysis performed using 1000 replicates. Two groups, namely group I and group II, are colored with blue and red, respectively, and the colored circles, rhombuses and squares represent different plant species.

3.3. Conserved Motifs of SRO Proteins from Cucumber and Other Plant Species

Domain analysis by Pfam and InterProScan tools showed that all CsSROs consist of the PARP domain and a C-terminal RST domain, while the N-terminal WWE domain was only present in CsSRO2 and CsSRO4 (Figure S2). To further investigate the diversification of SRO proteins, the conserved motif analysis of SRO proteins from cucumber and other plant species was performed based on the results of MEME, and a phylogenetic tree was also created. The results showed that 10 conversed motifs were predicted and their details are shown, and the SROs in the same group had similar motif composition (Figure 2). Motifs 9, 4, 8, 2, and 1 made up the PARP domain, which were widely distributed in a large proportion of SRO proteins, with the exception of several SROs, such as OsSRO1d and VvSRO1c (lack of motif 2), ZmSRO1d, ZmSRO1e, and AtSRO3 (lack of motif 9). Motifs 7 and 5 were annotated as the RST domain, which were present in most SRO proteins, except for CsSRO2, OsSRO1d, VvSRO1c, ZmSRO1d, and ZmSRO1e. In addition, motif 3 and motif 6 were present in the N-terminal region of all SRO proteins of group I, while they were absent in group II SRO proteins. Motif 10 was specific to SRO proteins of group Ia and Ib, as well as present in most members of group II, with the exception of AtSRO4 and AtSRO5 (Figure 2).

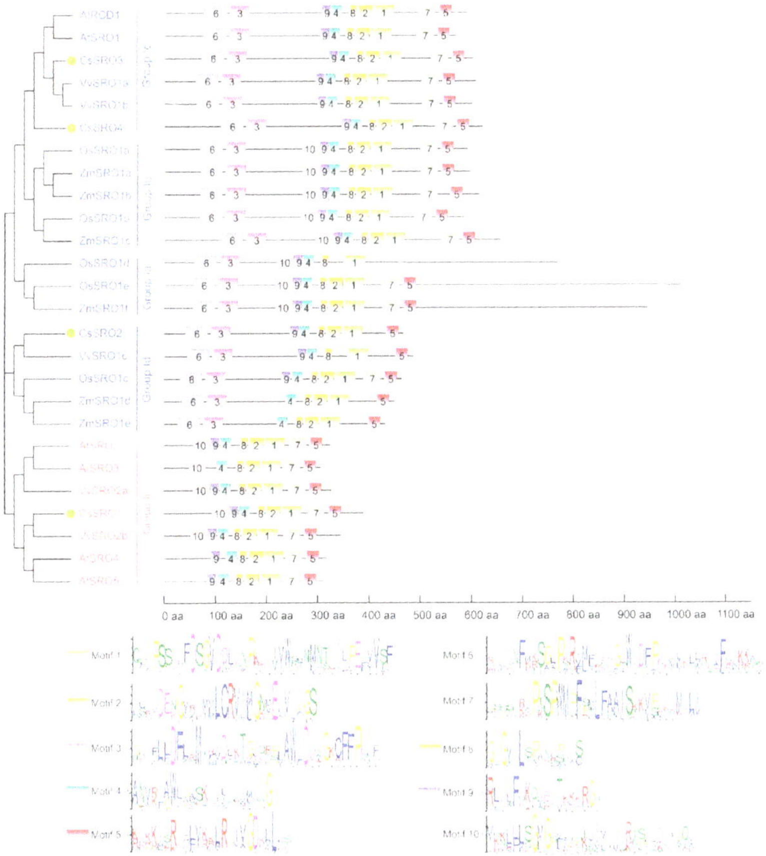

Figure 2. Conserved motif configurations of the *SRO* family members from cucumber, *Arabidopsis*, maize, rice, and grape on the basis of phylogenetic relationship. The clustering was shown according to the results of phylogenetic analysis. The colorful boxes delineate different motifs (numbered 1–10) and their details were shown at the bottom of the figure.

3.4. Gene Structure of SRO Family Genes

A detailed *SRO* structure analysis was performed by comparing the CDS sequences with the corresponding gDNA sequences of *SRO* family genes from cucumber, *Arabidopsis*, maize, rice, and grape. As a result, the number of introns in these *SRO* genes ranged from 3 to 7, and the structural characteristics of *SRO* genes clustered together were similar but varied among different groups (Figure 3). Notably, *SRO* genes in group Ia were found to have 4 or 7 introns, group Ib *SRO* genes had 5 or 6 introns, group Ic *SRO* genes harbored 4–7 introns, group Id *SRO* genes contained 3 introns each, while group II *SRO* genes were found to have 3 or 4 introns (Figure 3). The numbers of introns in four *CsSRO* genes were 4, 3, 5, 7, respectively.

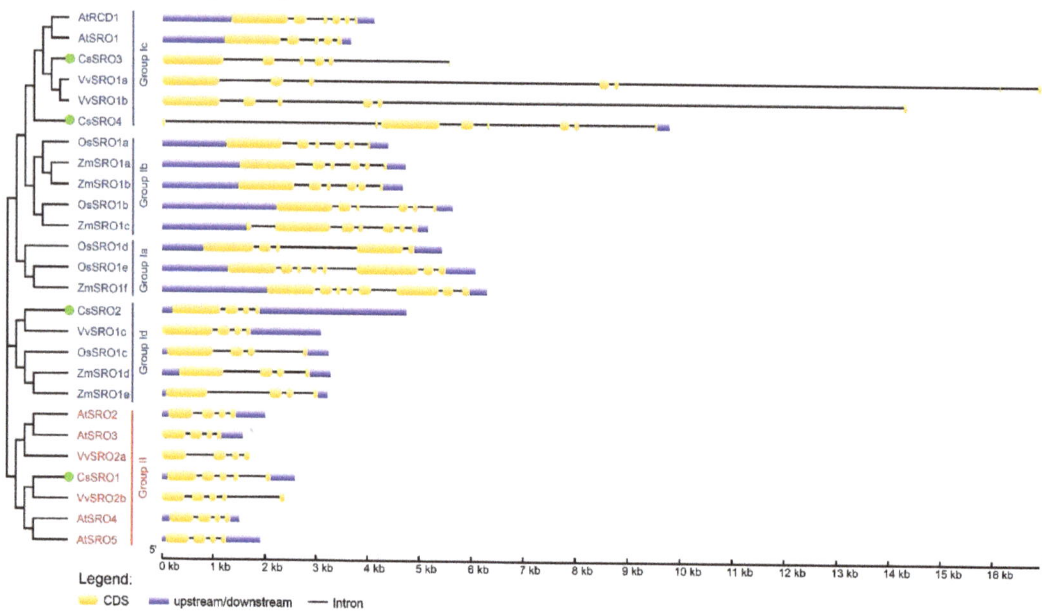

Figure 3. The gene structures of *SRO* family genes from cucumber, *Arabidopsis*, maize, rice, and grape on the basis of phylogenetic relationship. The clustering was shown based on the results of phylogenetic analysis. The boxes colored blue and yellow indicate upstream/downstream and CDS, respectively, the black line indicates intron.

3.5. The Distribution and Gene Duplication of CsSRO Genes

Chromosomal mapping of *CsSRO* genes showed that these genes were located on 3 out of 7 chromosomes of the cucumber genome. Chromosome 3 had the most *CsSRO* genes (two), whereas chromosomes 1 and 2 each possessed only one gene (Figure 4). The analysis of the duplication event indicated that a segmental duplication event was identified (*CsSRO3* and *CsSRO4*), while no tandem duplication event was detected (Figure 4).

3.6. Bioinformatics Analysis of Putative Promoters of CsSRO Genes

The *cis*-regulatory elements within the promoter regions of genes would provide valuable information for their expression and possible regulatory patterns. Therefore, the PlantCARE tool was employed to analyze the stress- and hormone-related *cis*-regulatory elements within the promoter regions of the *CsSRO* genes, and 16 kinds of *cis*-regulatory elements were identified (Figure 5). Eight kinds of *cis*-regulatory elements were found to be responsive to various hormones, such as ABRE for abscisic acid (ABA), TGACG-motif and CGTCA-motif for methyl jasmonate (MeJA), TGA-element for auxin, ERE for ethylene, TCA-element for salicylic acid (SA), P-box and TATC-box for gibberellin. It is worth noting that ABRE was most abundant in the promoter sequences of *CsSRO2*, implying that it may play a critical role in responsive to ABA. In addition, eight kinds of *cis*-regulatory elements are stress-responsive, including ARE (anaerobic induction), GC-motif (anoxia), LTR (low-temperature responsive), STRE (stress-responsive), W-box (fungal elicitors), TC-rich repeats (defense and stress response), WRE3 and WUN-motif (wounding responses) (Figure 5).

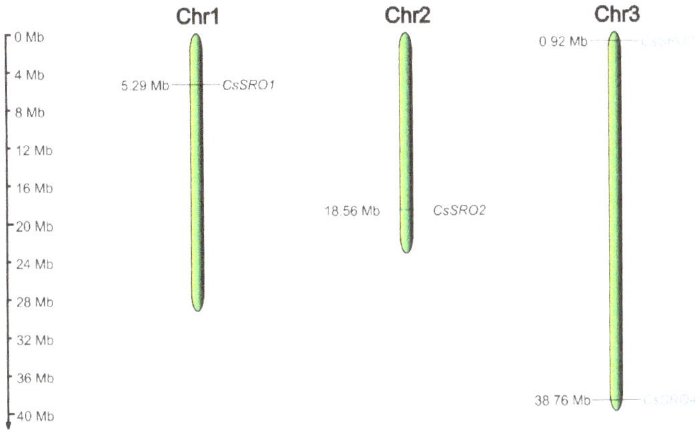

Figure 4. Distribution of *CsSRO* genes on cucumber chromosomes. The segmental duplication genes are colored with blue. The chromosome numbers are labeled on the top of the bars, and the chromosome length represents the size of chromosomes. Positions are in Mb.

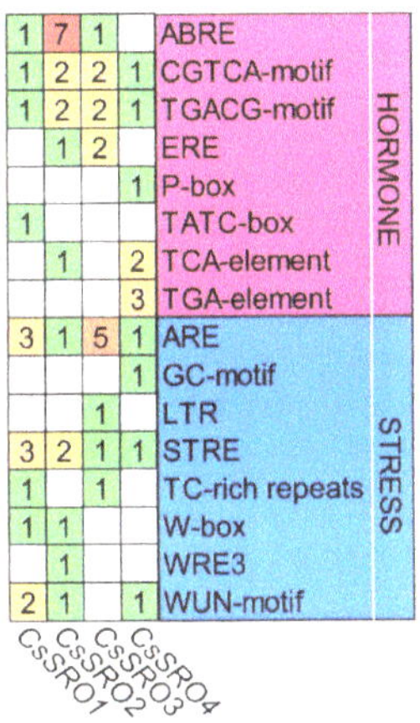

Figure 5. Stress- and hormone-responsive *cis*-regulatory elements in the promoter regions of *CsSRO* genes. The number in box represents the copy of the cis-element of individual *CsSRO* genes. ABRE, *cis*-element involved in the abscisic acid (ABA) responsiveness; CGTCA-motif and TGACG-motif, *cis*-element involved in the MeJA responsiveness; ERE, ethylene-responsive element; P-box and TATC-box, *cis*-elements involved in the gibberellin responsiveness; TCA-element, salicylic acid (SA)-responsive element; TGA-element, auxin-responsive element; ARE, *cis*-acting regulatory element essential for the anaerobic induction; GC-motif, anoxia-specific inducibility element; LTR, low-temperature responsive element; STRE, stress responsive element; W-box, WRKY binding site involved in fungal elicitors; TC-rich repeats, *cis*-acting element involved in defense and stress responsiveness, WRE3 and WUN-motif, wound-responsive elements.

3.7. Tissue Expression Profiles Analysis of CsSRO Genes

RNA-Seq data from different tissues (root, stem, leaf, flowers, and tendrils) and three stages of developing ovaries were used to examine the tissue-specific expression profiles of *CsSRO* genes, and an expression heat map was generated. As a result, *CsSRO* genes were universally expressed in all tested tissues, and *CsSRO3* and *CsSRO4* exhibited relatively higher expression levels than the other two *CsSRO* genes (Figure 6). *CsSRO1* displayed a much higher expression in root than in other tissues, and *CsSRO2* showed the most abundant gene expression levels in root, unfertilized ovary, male and female flowers (Figure 6).

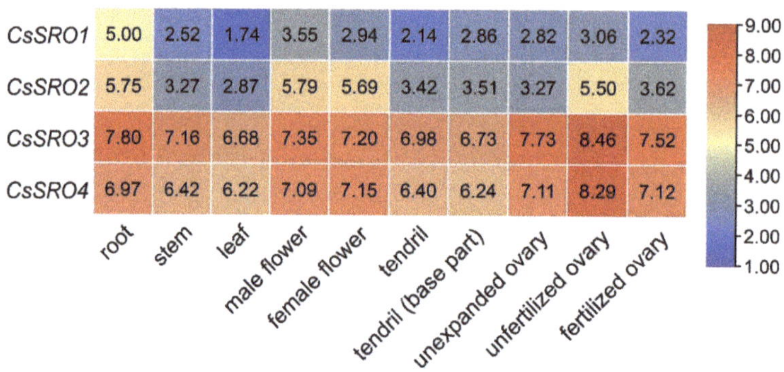

Figure 6. Expression profiles of the *CsSRO* genes in different tissues of cucumber based on RNA-Seq data. Data were normalized based on the $\log_2$-transformed TPM values of each gene in all tissues analyzed, and genes highly or weakly expressed are colored red or blue, respectively.

3.8. Expression Patterns of CsSRO Genes under Salt and Drought Stress

Since the promoter analysis revealed several stress-related *cis*-elements, we examined the expression patterns of *CsSRO* genes under salt and drought stress. Under salt stress, the expression of *CsSRO1, CsSRO3,* and *CsSRO4* was induced and peaked at 3 h, followed by an obvious decrease at 6 h and 12 h (Figure 7A). The expression level of *CsSRO2* was increased rapidly and then maintained at 6 h, and then dropped at 12 h. Under drought treatment, the *CsSRO* genes exhibited similar expression profiles as under salt treatment (Figure 7B). These results indicated that *CsSRO* genes may play a role in response to various abiotic stresses.

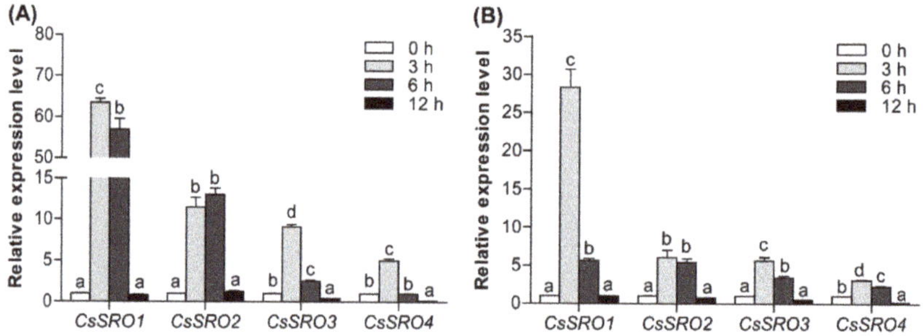

Figure 7. qRT-PCR analysis of *CsSRO* genes expression patterns under salt (**A**) and drought (**B**) stress. Cucumber seedlings were treated with salt (200 mM NaCl) and drought (10% PEG6000) conditions at different time points (0, 3, 6, and 12 h), and leaf samples were collected from three biological replicates of each treatment. Relative gene expression levels were normalized to *actin* expression levels, and the expression levels of the control (0 h) were set to 1.0. Error bars represent the standard deviations. The small letters above the bars indicate the significant differences between means ($p < 0.05$).

3.9. Expression Patterns of CsSRO Genes in Responses to Various Pathogen Infections

To analyze whether the *CsSRO* genes responded to biotic stress, their expression patterns under the infection of *P. cubensis* (DM treatment) and *M. incognita* (RKN treatment) were determined on the basis of the published RNA-Seq data [4,47]. In the Vlaspik (DM-susceptible) cucumber plants, *CsSRO1* displayed increased expression after *P. cubensis* infection at 2, 4, and 6 dpi compared to the control, while *CsSRO3* and *CsSRO4* were down-regulated at certain time points. In the PI 197088 (DM-resistant cucumber plant), *CsSRO1* displayed increased expression after *P. cubensis* infection at the earlier time point (1 dpi), while the other three *CsSRO* genes displayed decreased expression at certain time points, especially at 2 dpi (Figure 8A). Under RKN inoculation, the transcription levels of *CsSRO1* and *CsSRO2* were remarkably increased in susceptible CC3 lines, but no changed regulation of *CsSRO* genes was found during resistant IL10-1 lines response to RKN inoculation (Figure 8B).

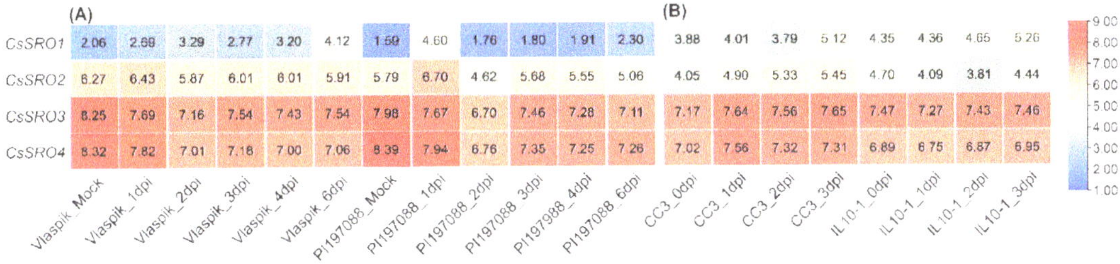

Figure 8. Expression profiles of *CsSRO* genes in response to DM (**A**) and RKN (**B**) based on RNA-Seq data. (**A**) Expression profiles of *CsSRO* genes in Vlaspik (DM-susceptible cucumber plant) and PI 197088 (DM-resistant cucumber plant) leaves that were mock inoculated or collected 1, 2, 3, 4, and 6 dpi with *P. cubensis* infection. (**B**) Expression profiles of *CsSRO* genes in root tips from CC3 (RKN susceptible cucumber plant) and IL10–1 (RKN resistant cucumber plant) at 0, 1, 2 and 3 dpi with *M. incognita* inoculation. Genes highly or weakly expressed are colored red or blue, respectively, based on the log_2 transformed TPM values.

4. Discussion

In recent decades, *SRO* gene family has been identified in diverse plant species whose genomes were sequenced [7,21,48]. However, no comprehensive identification and characterization has been performed for *SRO* genes in cucumber. In this study, the cucumber *SRO* family genes were identified and cloned, and these genes were named as *CsSRO1*–*CsSRO4*, making it a small gene family. The number of *SRO* family genes was compared to those of previously identified plant species, such as sesame, rice, *Arabidopsis*, maize, *Brachypodium distachyon*, tomato, banana, and apple, which generally contained 4–6 *SRO* genes in their genomes [46,48–50], but was much less than that in Chinese cabbage (12 genes) [51], and wheat (30 genes) [45]. These findings indicated that the expansion of the *SRO* gene family was relatively conservative, and the number of *SRO* family genes may not possess an absolute correlation with genome size during plant evolution. In addition, a segmental duplication event was found in this study (Figure 4), and similar results were also found in other plants, including wheat [45], and tomato [50].

Phylogenetic tree analysis showed that CsSRO proteins along with SRO proteins from 14 other different plant species were grouped into two groups (group I and group II) based on their sequence relationships, and group I can be further divided into four subgroups (Ia to Id) (Figure 1). The SRO proteins of these tested species were clustered in dicot- or monocot-specific patterns, and monocot species contained only group I SROs. In group I, monocotyledonous and dicotyledonous SRO proteins were also had individual distribution in special subgroups, of which group Ia and group Ib SROs were monocotyledonous, while SROs in group Ic and group Id were coming from mono- and dicot plants (Figure 1), suggesting that the evolution of SROs in these groups followed the monocot and dicot divergence. Notably, SROs including the WWE domain were distributed in group I, while all SROs in group II lack the WWE domain, which is consistent with the previous studies [7,48]. It can be speculated that the variation of the WWE domain is a divergent force for the expansion of the *SRO* gene family. We also identified 10 diverse conserved motifs, and there is an obvious difference in the arrangements of these conserved motifs in the two groups, while the SROs in the same group showed conservative conserved motif arrangements (Figure 2). Gene structure analysis provides extra clues into the procedure of evolution in different gene families [3]. In the current study, most *SRO* genes in the same subgroup displayed conservative gene structure, but they have variable numbers of introns in different subgroups (Figure 3), indicating that gain or loss of introns may present in *SRO* genes during the evolution, and the functions of phylogenetically related *SRO* genes may be conserved.

Previous reports revealed that the transcription of *SRO* genes exhibited constitutive expression or tissue-specific expression patterns. For example, *Arabidopsis AtSRO1* and *AtRCD1* [16], rice *OsSRO1d* and *OsSRO1e* [21], tomato *SolySRO5* and *SolySRO6* [50] were found to have a broad expression in various tissues. In sesame, *SiSRO1a* and *SiSRO1b* were highly transcribed in various tissues, while *SiSRO2a* and *SiSRO2b* exhibited the highest transcript levels in root [46]. We thus investigated the spatio-temporal expression patterns of *CsSRO* genes during the development of various tissues based on the published RNA-Seq data. The results showed that *CsSRO1* was expressed at a much higher level in root than other tested tissues, *CsSRO2* was highly expressed in root, unfertilized ovary, female and male flowers, while *CsSRO3* and *CsSRO4* were ubiquitously expressed in the tested tissues (Figure 6). In apple, the expression of *MdSRO1* and *MdSRO4* was highest in roots, whereas *MdSRO5* displayed the highest expression in flowers [52]. In tomato, the highest transcript levels of *SolySRO2*, *SolySRO3*, and *SolySRO5* were observed in flowers, roots, and 3 cm fruits, respectively [50]. In addition, *CsSRO2* and *CsSRO4* displayed obviously higher transcription levels in unfertilized ovary than unexpanded and fertilized ovaries (Figure 6), thus fueling our speculation that they may participate in ovary development of cucumber. Similarly, *SolySRO4* was highly expressed in mature tomato fruits (breaker + 10 fruits), and its values were much higher than that in mature green fruits and breaker fruits, suggesting that this gene may participate in fruit development and ripening [50]. Moreover, the identified segmental duplication gene pairs, *CsSRO3* and *CsSRO4*, showed very similar spatio-temporal expression patterns (Figure 6), suggesting their possible similar functions.

It was reported that genes in *SRO* family play vital roles in the resistance to various abiotic stresses. In this work, many *cis*-elements responding to stresses (especially ARE and STRE) were distributed in the promoters of *CsSRO* genes (Figure 5), and similar results were reported in the promoter regions of *SolySRO* and *SiSRO* genes [46,50]. The qRT-PCR analysis also revealed that the expression of four *CsSRO* genes were significantly induced under salt and drought stress (Figure 7), suggesting their involvement in stress tolerance. In previous reports, rice *SRO* genes (*OsSRO1a–OsSRO1e*) were induced by a series of abiotic stresses, including heat, cold, drought, UV, salt, wound, and submergence, and *OsSRO1c* was found to play positive roles in cold, drought, and oxidative stress [21,22]. Transgenic *Arabidopsis* plants overexpressing *IcSRO1* from *Ipomoea cairica* also exhibited enhanced tolerance to salt and drought stress [53]. Overexpression of *MdRCD1* in *Arabidopsis* promotes plant adaptation to drought, salt, and H_2O_2 stresses [52]. Overexpression of *SiSRO2b* also conferred the tolerance of transgenic yeast cells to osmotic, salt, and oxidative stress [46]. *SRO* family genes are also known as regulators function in plant defense response against pathogens, and the regulatory mechanisms depend on the hormone signaling pathways, such as JA, SA, and so on. For example, OsSRO1a was found to regulate JA-dependent disease susceptibility to the *Xanthomonas oryzae* pv *oryzae* (*Xoo*) [24]. The presence of *cis*-elements associated with hormones and defense responses in the promoters of *CsSRO* genes implied the potential roles of *CsSRO* genes in response to biotic stress through the hormone signal transduction pathway (Figure 5). Additionally, all *CsSRO* genes showed changed expression levels in both sensitive and resistant cucumber plants inoculated with DM, except for *CsSRO2*, whose expression was unchanged in DM-susceptible Vlaspik plants (Figure 8). *CsSRO1* and *CsSRO2* displayed significantly increased expression at certain time points in susceptible CC3 plants inoculated with RKN (Figure 8). In banana, a total of four out of six *MaSRO* genes showed increased expression levels during *Fusarium oxysporum* f. sp. *Cubense* tropical race 4 (*Foc* TR4) infection [48]. Therefore, *CsSRO* genes may also play important roles in the response of plants to adverse environmental conditions.

5. Conclusions

In this work, four *SRO* family genes were isolated from cucumber, and their evolutionary relationship, conserved motif composition, gene structure, chromosomal location, *cis*-regulatory elements within the promoter region, followed by expression patterns in various tissues and differential expression in response to abiotic and biotic stresses were systematically analyzed. Expression analysis showed that the four *CsSRO* genes were differentially expressed in various tissues and under diverse stresses, suggesting their roles in plant physiological processes and defense response. Our findings facilitate further study to the functional characterization of *CsSRO* genes in response to abiotic and biotic stresses in cucumber, and these *CsSRO* genes can be considered as potential candidates for breeding of cucumber stress-tolerant varieties in the future.

Supplementary Materials: The following are available online at https://www.mdpi.com/article/10.3390/horticulturae8070634/s1, Figure S1: PCR amplification of *CsSRO* genes. M, DL ladder 2000 DNA Marker; 1, *CsSRO1*; 2, *CsSRO2*; 3, *CsSRO3*; 4, *CsSRO4*, Figure S2: Schematic diagram of domain distributions in CsSRO proteins; Table S1: The gene-specific primers used in this study.

Author Contributions: Data curation, L.X., Z.Z., C.Z. and Y.Z.; formal analysis, C.Z. and J.Z.; funding acquisition, S.L. and Y.Z.; investigation, L.X., Z.Z. and J.Z.; methodology, L.X., Z.Z., C.Z., J.Z., Z.H. and Y.Z.; resources, L.X., Z.Z., Z.H. and S.L.; software, C.Z. and Y.Z.; writing—original draft, L.X. and Y.Z.; writing—review and editing, S.L. and Y.Z. All authors have read and agreed to the published version of the manuscript.

Funding: This work was funded by the Academic and Technical Leader Plan of Jiangxi Provincial Main Disciplines (20204BCJ22023), the Natural Science Foundation of Jiangxi Province (20202BABL205002), and the National Natural Science Foundation of China (31860566).

Institutional Review Board Statement: Not applicable.

Informed Consent Statement: Not applicable.

Data Availability Statement: The data presented in this study are available within the article and Supplementary Materials.

Conflicts of Interest: The authors declare that they have no conflict of interest.

References

1. Dos Santos, T.B.; Ribas, A.F.; de Souza, S.G.H.; Budzinski, I.G.F.; Domingues, D.S. Physiological responses to drought, salinity, and heat stress in plants: A review. *Stresses* **2022**, *2*, 113–135. [CrossRef]
2. Kalia, V.C.; Gong, C.; Patel, S.K.S.; Lee, J.K. Regulation of plant mineral nutrition by signal molecules. *Microorganisms* **2021**, *9*, 774. [CrossRef] [PubMed]
3. Zhou, Y.; Li, G.; Zhang, L.; Xu, J.; Hu, L.; Jiang, L.; Liu, S. Comprehensive genomic analysis and expression profiling of the BTB and TAZ (*BT*) genes in cucumber (*Cucumis sativus* L.). *Czech J. Genet. Plant Breed.* **2020**, *56*, 15–23. [CrossRef]
4. Zhu, C.; Xiao, L.; Hu, Y.; Liu, L.; Liu, H.; Hu, Z.; Liu, S.; Zhou, Y. Genome-wide survey and expression analysis of B-box family genes in cucumber reveal their potential roles in response to diverse abiotic and biotic stresses. *Agriculture* **2022**, *12*, 827. [CrossRef]
5. Aravind, L. The WWE domain: A common interaction module in protein ubiquitination and ADP ribosylation. *Trends Biochem. Sci.* **2001**, *26*, 273–275. [CrossRef]
6. Jaspers, P.; Brosché, M.; Overmyer, K.; Kangasjärvi, J. The transcription factor interacting protein RCD1 contains a novel conserved domain. *Plant Signal. Behav.* **2010**, *5*, 78–80. [CrossRef]
7. Jaspers, P.; Overmyer, K.; Wrzaczek, M.; Vainonen, J.P.; Blomster, T.; Salojärvi, J.; Reddy, R.A.; Kangasjärvi, J. The RST and PARP-like domain containing SRO protein family: Analysis of protein structure, function and conservation in land plants. *BMC Genom.* **2010**, *11*, 170. [CrossRef]

8. Belles-Boix, E.; Babiychuk, E.; Van Montagu, M.; Inzé, D.; Kushnir, S. CEO1, a new protein from *Arabidopsis thaliana*, protects yeast against oxidative damage. *FEBS Lett.* **2000**, *482*, 19–24. [CrossRef]
9. Katiyar-Agarwal, S.; Zhu, J.; Kim, K.; Agarwal, M.; Fu, X.; Huang, A.; Zhu, J.K. The plasma membrane Na^+/H^+ antiporter SOS1 interacts with RCD1 and functions in oxidative stress tolerance in *Arabidopsis*. *Proc. Natl. Acad. Sci. USA* **2006**, *103*, 18816–18821. [CrossRef] [PubMed]
10. Overmyer, K.; Tuominen, H.; Kettunen, R.; Betz, C.; Langebartels, C.; Sandermann, H., Jr.; Kangasjärvi, J. Ozone-sensitive arabidopsis rcd1 mutant reveals opposite roles for ethylene and jasmonate signaling pathways in regulating superoxide-dependent cell death. *Plant Cell* **2000**, *12*, 1849–1862. [CrossRef]
11. Ahlfors, R.; Lång, S.; Overmyer, K.; Jaspers, P.; Brosché, M.; Tauriainen, A.; Kollist, H.; Tuominen, H.; Belles-Boix, E.; Piippo, M.; et al. Arabidopsis RADICAL-INDUCED CELL DEATH1 belongs to the WWE protein-protein interaction domain protein family and modulates abscisic acid, ethylene, and methyl jasmonate responses. *Plant Cell* **2004**, *16*, 1925–1937. [CrossRef] [PubMed]
12. Fujibe, T.; Saji, H.; Arakawa, K.; Yabe, N.; Takeuchi, Y.; Yamamoto, K.T. A methyl viologen-resistant mutant of *Arabidopsis*, which is allelic to ozone-sensitive *rcd1*, is tolerant to supplemental ultraviolet-B irradiation. *Plant Physiol.* **2004**, *134*, 275–285. [CrossRef] [PubMed]
13. Teotia, S.; Lamb, R.S. The paralogous genes *RADICAL-INDUCED CELL DEATH1* and *SIMILAR TO RCD ONE1* have partially redundant functions during *Arabidopsis* development. *Plant Physiol.* **2009**, *151*, 180–198. [CrossRef]
14. Teotia, S.; Lamb, R.S. RCD1 and SRO1 are necessary to maintain meristematic fate in *Arabidopsis thaliana*. *J. Exp. Bot.* **2011**, *62*, 1271–1284. [CrossRef]
15. Vainonen, J.P.; Jaspers, P.; Wrzaczek, M.; Lamminmäki, A.; Reddy, R.A.; Vaahtera, L.; Brosché, M.; Kangasjärvi, J. RCD1-DREB2A interaction in leaf senescence and stress responses in *Arabidopsis thaliana*. *Biochem. J.* **2012**, *442*, 573–581. [CrossRef]
16. Jaspers, P.; Blomster, T.; Brosché, M.; Salojärvi, J.; Ahlfors, R.; Vainonen, J.P.; Reddy, R.A.; Immink, R.; Angenent, G.; Turck, F.; et al. Unequally redundant RCD1 and SRO1 mediate stress and developmental responses and interact with transcription factors. *Plant J.* **2009**, *60*, 268–279. [CrossRef] [PubMed]
17. Jiang, L.; Wang, Y.; Björn, L.O.; Li, S. Arabidopsis radical-induced cell death1 is involved in UV-B signaling. *Photochem. Photobiol. Sci.* **2009**, *8*, 838–846. [CrossRef]
18. Li, B.Z.; Zhao, X.; Zhao, X.L.; Peng, L. Structure and function analysis of *Arabidopsis thaliana* SRO protein family. *Yi Chuan* **2013**, *35*, 1189–1197. [CrossRef]
19. Zhao, X.; Gao, L.; Ren, J.; Pan, F. *Arabidopsis SIMILAR TO RCD-ONE* genes are ubiquitous and respond to multiple abiotic stresses through diverse signaling pathways. *J. Biosci.* **2019**, *44*, 129. [CrossRef]
20. Borsani, O.; Zhu, J.; Verslues, P.E.; Sunkar, R.; Zhu, J.K. Endogenous siRNAs derived from a pair of natural cis-antisense transcripts regulate salt tolerance in *Arabidopsis*. *Cell* **2005**, *123*, 1279–1291. [CrossRef]
21. You, J.; Zong, W.; Du, H.; Hu, H.; Xiong, L. A special member of the rice SRO family, OsSRO1c, mediates responses to multiple abiotic stresses through interaction with various transcription factors. *Plant Mol. Biol.* **2014**, *84*, 693–705. [CrossRef] [PubMed]
22. You, J.; Zong, W.; Li, X.; Ning, J.; Hu, H.; Li, X.; Xiao, J.; Xiong, L. The SNAC1-targeted gene OsSRO1c modulates stomatal closure and oxidative stress tolerance by regulating hydrogen peroxide in rice. *J. Exp. Bot.* **2013**, *64*, 569–583. [CrossRef]
23. Zhang, K.; Su, J.; Xu, M.; Zhou, Z.; Zhu, X.; Ma, X.; Hou, J.; Tan, L.; Zhu, Z.; Cai, H.; et al. A common wild rice-derived *BOC1* allele reduces callus browning in *indica* rice transformation. *Nat. Commun.* **2020**, *11*, 443. [CrossRef] [PubMed]
24. Kashihara, K.; Onohata, T.; Yariuchi, R.; Tanaka, S.; Akimitsu, K.; Gomi, K. The overexpression of OsSRO1a, which encodes an OsNINJA1- and OsMYC2-interacting protein, negatively affects OsMYC2-mediated jasmonate signaling in rice. *Plant Cell Rep.* **2020**, *39*, 489–500. [CrossRef] [PubMed]
25. Liu, S.; Liu, S.; Wang, M.; Wei, T.; Meng, C.; Wang, M.; Xia, G. A wheat SIMILAR TO RCD-ONE gene enhances seedling growth and abiotic stress resistance by modulating redox homeostasis and maintaining genomic integrity. *Plant Cell* **2014**, *26*, 164–180. [CrossRef]
26. Li, X.; Xu, Y.; Liu, F.; Zhao, M.; Sun, Y.; Ma, Q. Maize similar to RCD1 gene induced by salt enhances *Arabidopsis thaliana* abiotic stress resistance. *Biochem. Biophys. Res. Commun.* **2018**, *503*, 2625–2632. [CrossRef]
27. Qin, L.; Sun, L.; Wei, L.; Yuan, J.; Kong, F.; Zhang, Y.; Miao, X.; Xia, G.; Liu, S. Maize SRO1e represses anthocyanin synthesis through regulating the MBW complex in response to abiotic stress. *Plant J.* **2021**, *105*, 1010–1025. [CrossRef]
28. Liao, L.; Hu, Z.; Liu, S.; Yang, Y.; Zhou, Y. Characterization of germin-like proteins (GLPs) and their expression in response to abiotic and biotic stresses in cucumber. *Horticulturae* **2021**, *7*, 412. [CrossRef]
29. Lu, X.; Liu, W.; Xiang, C.; Li, X.; Wang, Q.; Wang, T.; Liu, Z.; Zhang, J.; Gao, L.; Zhang, W. Genome-wide characterization of GRAS family and their potential roles in cold tolerance of cucumber (*Cucumis sativus* L.). *Int. J. Mol. Sci.* **2020**, *21*, 3857. [CrossRef] [PubMed]
30. Li, C.; Dong, S.; Liu, X.; Bo, K.; Miao, H.; Beckles, D.M.; Zhang, S.; Gu, X. Genome-wide characterization of cucumber (*Cucumis sativus* L.) GRAS genes and their response to various abiotic stresses. *Horticulturae* **2020**, *6*, 110. [CrossRef]
31. Li, J.; Wang, T.; Han, J.; Ren, Z. Genome-wide identification and characterization of cucumber bHLH family genes and the functional characterization of CsbHLH041 in NaCl and ABA tolerance in *Arabidopsis* and cucumber. *BMC Plant Biol.* **2020**, *20*, 272. [CrossRef] [PubMed]

32. Chen, C.; Chen, X.; Han, J.; Lu, W.; Ren, Z. Genome-wide analysis of the WRKY gene family in the cucumber genome and transcriptome-wide identification of WRKY transcription factors that respond to biotic and abiotic stresses. *BMC Plant Biol.* **2020**, *20*, 443. [CrossRef] [PubMed]
33. Lai, W.; Zhu, C.; Yang, S.; Hu, Z.; Liu, S.; Zhou, Y. Comprehensive identification of the *VQ* family genes in cucumber and their roles in response to abiotic and biotic stresses. *Sci. Hortic.* **2022**, *295*, 110874. [CrossRef]
34. Dai, Z.; Dong, S.; Miao, H.; Liu, X.; Han, J.; Li, C.; Gu, X.; Zhang, S. Genome-wide identification of TIFY genes and their response to various pathogen infections in cucumber (*Cucumis sativus* L.). *Sci. Hortic.* **2022**, *295*, 110814. [CrossRef]
35. Luan, Q.; Chen, C.; Liu, M.; Li, Q.; Wang, L.; Ren, Z. CsWRKY50 mediates defense responses to *Pseudoperonospora cubensis* infection in *Cucumis sativus*. *Plant Sci.* **2019**, *279*, 59–69. [CrossRef]
36. Zhang, Y.; Yu, H.; Yang, X.; Li, Q.; Ling, J.; Wang, H.; Gu, X.; Huang, S.; Jiang, W. CsWRKY46, a WRKY transcription factor from cucumber, confers cold resistance in transgenic-plant by regulating a set of cold-stress responsive genes in an ABA-dependent manner. *Plant Physiol. Biochem.* **2016**, *108*, 478–487. [CrossRef]
37. El-Gebali, S.; Mistry, J.; Bateman, A.; Eddy, S.R.; Luciani, A.; Potter, S.C.; Qureshi, M.; Richardson, L.J.; Salazar, G.A.; Smart, A.; et al. The Pfam protein families database in 2019. *Nucleic. Acids Res.* **2019**, *47*, D427–D432. [CrossRef]
38. Gasteiger, E.; Gattiker, A.; Hoogland, C.; Ivanyi, I.; Appel, R.D.; Bairoch, A. ExPASy: The proteomics server for in-depth protein knowledge and analysis. *Nucleic Acids Res.* **2003**, *31*, 3784–3788. [CrossRef]
39. Yu, C.S.; Chen, Y.C.; Lu, C.H.; Hwang, J.K. Prediction of protein subcellular localization. *Proteins* **2006**, *64*, 643–651. [CrossRef]
40. Lescot, M.; Déhais, P.; Thijs, G.; Marchal, K.; Moreau, Y.; Van de Peer, Y.; Rouzé, P.; Rombauts, S. PlantCARE, a database of plant *cis*-acting regulatory elements and a portal to tools for *in silico* analysis of promoter sequences. *Nucleic Acids Res.* **2002**, *30*, 325–327. [CrossRef]
41. Kumar, S.; Stecher, G.; Tamura, K. MEGA7: Molecular evolutionary genetics analysis version 7.0 for bigger datasets. *Mol. Biol. Evol.* **2016**, *33*, 1870–1874. [CrossRef] [PubMed]
42. Bailey, T.L.; Boden, M.; Buske, F.A.; Frith, M.; Grant, C.E.; Clementi, L.; Ren, J.; Li, W.W.; Noble, W.S. MEME SUITE: Tools for motif discovery and searching. *Nucleic Acids Res.* **2009**, *37*, W202–W208. [CrossRef]
43. Hu, B.; Jin, J.; Guo, A.Y.; Zhang, H.; Luo, J.; Gao, G. GSDS 2.0: An upgraded gene feature visualization server. *Bioinformatics* **2015**, *31*, 1296–1297. [CrossRef] [PubMed]
44. Lai, W.; Zhu, C.; Hu, Z.; Liu, S.; Wu, H.; Zhou, Y. Identification and transcriptional analysis of zinc finger-homeodomain (ZF-HD) family genes in cucumber. *Biochem. Genet.* **2021**, *59*, 884–901. [CrossRef]
45. Jiang, W.; Geng, Y.; Liu, Y.; Chen, S.; Cao, S.; Li, W.; Chen, H.; Ma, D.; Yin, J. Genome-wide identification and characterization of SRO gene family in wheat: Molecular evolution and expression profiles during different stresses. *Plant Physiol. Biochem.* **2020**, *154*, 590–611. [CrossRef]
46. Liu, A.; Wei, M.; Zhou, Y.; Li, D.; Zhou, R.; Zhang, Y.; Zhang, X.; Wang, L.; You, J. Comprehensive analysis of *SRO* gene family in *Sesamum indicum* (L.) reveals its association with abiotic stress responses. *Int. J. Mol. Sci.* **2021**, *22*, 13048. [CrossRef] [PubMed]
47. Wang, X.; Cheng, C.; Zhang, K.; Tian, Z.; Xu, J.; Yang, S.; Lou, Q.; Li, J.; Chen, J.F. Comparative transcriptomics reveals suppressed expression of genes related to auxin and the cell cycle contributes to the resistance of cucumber against *Meloidogyne incognita*. *BMC Genom.* **2018**, *19*, 583. [CrossRef] [PubMed]
48. Zhang, L.; Zhou, D.; Hu, H.; Li, W.; Hu, Y.; Xie, J.; Huang, S.; Wang, W. Genome-wide characterization of a *SRO* gene family involved in response to biotic and abiotic stresses in banana (*Musa* spp.). *BMC Plant Biol.* **2019**, *19*, 211. [CrossRef]
49. Jiang, H.; Xiao, Y.; Zhu, S. Genome-wide identification, systematic analysis and characterization of *SRO* family genes in maize (*Zea mays* L.). *Acta Physiol. Plant* **2018**, *40*, 176. [CrossRef]
50. Li, N.; Xu, R.; Wang, B.; Wang, J.; Huang, S.; Yu, Q.; Gao, J. Genome-wide identification and evolutionary analysis of the SRO gene family in tomato. *Front. Genet.* **2021**, *12*, 753638. [CrossRef]
51. Qiao, Y.; Gao, X.; Liu, Z.; Wu, Y.; Hu, L.; Yu, J. Genome-wide identification and analysis of *SRO* gene family in Chinese cabbage (*Brassica rapa* L.). *Plants* **2020**, *9*, 1235. [CrossRef] [PubMed]
52. Li, H.; Li, R.; Qu, F.; Yao, J.; Hao, Y.; Wang, X.; You, C. Identification of the *SRO* gene family in apples (*Malus* × *domestica*) with a functional characterization of *MdRCD1*. *Tree Genet. Genomes* **2017**, *13*, 94. [CrossRef]
53. Yuan, B.; Chen, M.; Li, S. Isolation and Identification of *Ipomoea cairica* (L.) sweet gene *IcSRO1* encoding a SIMILAR TO RCD-ONE protein, which improves salt and drought tolerance in transgenic *Arabidopsis*. *Int. J. Mol. Sci.* **2020**, *21*, 1017. [CrossRef] [PubMed]

horticulturae

MDPI

Article

Physiological and Transcriptomic Analyses Reveal the Response of Medicinal Plant *Bletilla striata* (Thunb. ex A. Murray) Rchb. f. via Regulating Genes Involved in the ABA Signaling Pathway, Photosynthesis, and ROS Scavenging under Drought Stress

Hai Liu [1], Kaizhang Chen [2], Lin Yang [1], Xue Han [1], Mingkai Wu [1,*] and Zhijun Shen [3,*]

[1] Institute of Modern Chinese Herbal Medicines, Guizhou Academy of Agricultural Sciences, Guiyang 550025, China
[2] Science and Technology Bureau of Pengyang County, Guyuan 756500, China
[3] Fujian Key Laboratory of Subtropical Plant Physiology and Biochemistry, Fujian Institute of Subtropical Botany, Xiamen 361006, China
* Correspondence: bywmk1999@163.com (M.W.); shenzj1987@xmu.edu.cn (Z.S.)

Abstract: *Bletilla striata* is a valuable Chinese herbal medicinal plant widely used in various fields. To meet the market demand for this herb, the tissue culture technology of *B. striata* was developed. However, drought stress has been a significant threat to the survival of cultivated *B. striata*. To further understand the underlying mechanisms of *B. striata* under drought stress, its response was investigated at the physiological and transcriptional levels. Our photosynthesis results show that the decline of the net photosynthesis rate (*Pn*) in *B. striata* leaves was mainly caused by nonstomatal limitation factors. Using transcriptomic analysis 2398, differentially expressed genes (DEGs) were identified. KEGG enrichment analysis showed that DEGs involved in plant hormone signal transduction (ko04075) were significantly altered, especially the abscisic-acid signaling pathway. The up-regulations of the serine/threonine protein kinase (*SnRK2*) and S-type anion (*SLAH2*) channels might lead to stomatal closure, which is the reason for decline of photosynthesis. Moreover, the downregulation of cytochrome b6 and photosystem I reaction center subunit III/IV might be the major reason for nonstomatal limitation. In addition, *B. striata* enhanced the ability of ROS scavenging via activating the gene expression of superoxide dismutase, catalase, and peroxidase in response to drought stress. Our study enhanced the understanding of *B. striata* in response to drought stress.

Keywords: *Bletilla striata*; medicinal plant; drought; transcriptome; ABA

1. Introduction

Drought is a critical environmental factor that seriously threatens plant growth and development [1]. Drought stress hampers various biological processes in plants, such as photosynthesis, cell elongation, nutrient uptake, and reproduction [2–6]. Plants have evolved a battery of defense mechanisms to ensure their survival and fitness in the face of a drought.

One plant response to drought is to reduce stomatal apertures so that leaf water potential is maintained [7]. While it restricts water from exiting, stomatal closure also restricts CO_2 from entering, thereby reducing the photosynthesis rate. Reactive oxygen species (ROSs) are also produced from photosynthesis and photorespiration [8]. The restriction of the photosynthesis rate caused by stomatal closure leads to overproduction of toxic ROS, which disrupts the electron transport system and starves cellular organelles of their carbon feedstock [7]. These toxic compounds cause oxidative damage that limits plant growth and development [8]. In response to this, a cell would synthesize antioxidants such as catalase (CAT), superoxide dismutase (SOD), peroxidase (POX), ascorbate peroxidase (APX), glutathione (GSH), and glutathione sulfo-transferase (GST) [9]. The biosyntheses of these complex antioxidants form the primary defense line against drought stress [10–12].

As a valuable Chinese herbal medicine, *Bletilla striata* (Thunb. ex A. Murray) Rchb. f. has received increasing attention from scholars in the last decade. The *Bletilla striata* polysccharide (BSP) is the important component in the *B. striata* tuber, purported to accelerate localized hemostasis in the lungs and stomach [13,14]. The development of omics technology enabled us to further explore the potential application value of *B. striata*. With bioinformatic analyses of the transcriptome data, auxin/indole-3-acetic acid (*Aux/IAA*) genes were suggested to participate in *B. striata* tuber development [15]. Similarly, another transcriptomic study proposed a pathway for BSP biosynthesis [16]. Recently, the genome of *B. striata* was sequenced and analyzed, and it will enhance molecular marker-assisted breeding of *B. striata* to improve traits of medicinal value [17]. These studies provide a solid foundation for *B. striata* planting and engineering of breeding for the future.

Although *B. striata* is widely distributed in China, the current industry of foraging *B. striata* from the wilderness is unable to meet the market demand not only due to its long growth periods and low breeding efficiency but also because it is widely used in various fields [18,19]. To solve this problem, the tissue culture technique was applied to successfully breed tissue culture seedlings [20]. However, the seedlings suffered from dehydration when they were transplanted to greenhouses and fields. Recently, a study found that PYLs-PP2C22/38-SnRK2s function as the ABA core signal pathway in response to multiple abiotic stresses [21]. Another study of *B. striata* suggested that an appropriate drought condition could improve its growth [22]. However, the mechanism of *B. striata* in response to drought stress is not well-known. To better understand the underlying mechanism of the *B. striata* response to drought stress, physiological and transcriptomic analyses were performed in the present study.

2. Materials and Methods

2.1. Plant Growth and Drought Treatment

Bletilla striata seedlings were produced through tissue culturing at the Institute of Modern Chinese Herbal Medicines, Guizhou Academy of Agricultural Sciences (Guizhou, China). Three seedlings with similar sizes (22–26 cm height, with 3 leaves) were transplanted into one pot (15 cm height × 14 cm diameter). A mixture of humus, vermiculite, and sand (1:1:1, *v:v:v*) was used as a soil matrix. Every twelve pots were put in a box (each group contained 36 seedlings at least). The seedlings were grown in a greenhouse at a daily temperature of 26–28 °C, a light intensity of 500–800 $\mu mL \cdot m^{-2} \cdot s^{-1}$, a photoperiod of 12/12 h (day/night), and a relative humidity of 50–80%. After two weeks of cultivation in the greenhouse, healthy seedlings were randomly divided into two groups (CK and drought).

At the beginning of the drought treatment, the plot containing soil matric was fully irrigated with tap water and put in a greenhouse until no water came out from the bottom of the plot. Then, the weight of the soil was recorded (g_1). The soil was dried out until its weight was stable (g_2). Subsequently, soil water content was calculated with the following formula: soil water content = $(g_1 - g_2)/g_2 \times 100\%$. Afterward, a certain amount of water (g_3) was added to the soil (soil water content = $g_3/(g_2 + g_3) \times 100\%$). Then, the *B. striate* seedlings were transplanted into these plots. The seedlings were irrigated with tap water every three days to maintain the soil water content at 25–35% (CK) or 5–10% (drought) for four weeks. Soil water content was real-time-monitored using time-domain reflectometry (Field Scout TDR 100, Spectrum Technologies Inc., Aurora, CO, USA). Each group had three replicates.

2.2. Leaf Photosynthesis Measurement

The second fully expanded leaf from the top was selected for photosynthesis measurement using the portable Li-6400XT photosynthesis measurement system (Li-6400, Li-Cor,

Lincoln, NE, USA). This test measured the net photosynthetic rate (*Pn*), intercellular CO_2 concentration (Ci), stomatal conductance (*Gs*), and transpiration rate (*E*) of the plant. From there, the stomatal limitation (*Ls*) and nonstomatal limitation (Ci/*Gs*) were calculated using formulae described in previous studies [23,24]. For each treatment, the photosynthesis measurement was performed on at least five seedlings. The photosynthetically active radiation (PAR) was set at 500 $\mu mol \cdot m^{-2} \cdot s^{-1}$, while the concentration of atmospheric CO_2 was maintained at around 400 $\mu mol \cdot mol^{-1}$.

2.3. RNA Extraction, Library Construction, and Sequencing

A 0.2 g sample of a fresh leaf was ground into powder in liquid nitrogen, and the total RNA was extracted using the Trizol (Invitrogen, Carlsbad, CA, USA) method, following the instructions provided by the vendor. The RNA quality was analyzed using the RNA Nano 6000 Assay Kit of the Bioanalyzer 2100 system (Agilent Technologies, Santa Clara, CA, USA). The value of the RIN (RNA integrity number) should have been higher than 8.5, and 28S:18S should have been higher than 1.5. After RNA extraction, a cDNA library was constructed according to vendor recommendation (Illumina, San Diego, CA, USA). The total RNA was used to isolate poly-(A) mRNA with Oligo-(dT) magnetic beads. The poly-(A) mRNA was fragmented in a fragmentation buffer. M-MuLV reverse transcriptase (RNase H-) was used for first-strand cDNA synthesis. The synthesis of second-strand cDNA was catalyzed with DNA polymerase in a buffer containing dNTPs and RNaseH. The passivation of the remaining overhangs was conducted through exonuclease/polymerase activities. With adenylation of 3′ ends of DNA fragments, a NEBNext adaptor with a hairpin loop structure was ligated to prepare for hybridization. The cDNA fragments with 250–300 bp lengths were purified with an AMPure XP system (Beckman Coulter, Beverly, CA, USA). Then, 3 μL of the USER enzyme (NEB, Ipswich, MA, USA) was used with size-selected, adaptor-ligated cDNA at 37 °C for 15 min. This selected cDNA was amplified by the polymerase chain reaction (PCR) in the mixed solution containing Universal PCR primers, Q5 Hot Start HiFi DNA polymerase, and Index (X) primer.

The amplified DNA were purified (AMPure XP system), and the quality of the cDNA library was assessed using the Agilent Bioanalyzer 2100 system. The index-coded samples were clustered using a cBot Cluster Generation System and a TruSeq PE Cluster Kit v4-cBot-HS (Illumia, San Diego, CA, USA) according to the vendor instructions. Afterward, the library preparations were sequenced on an Illumina Hiseq 2500 platform, and 100/50 bp single-end reads were generated.

2.4. De Novo Assembly and Sequence Annotation

Raw data were initially processed through in-house perl scripts, resulting in clean data, which was defined as having a Q20 value higher than 90% per our quality assessment criteria. The clean data were assembled using Trinity software [25]. Unigenes were generated and aligned to various databases. The thresholds for the Nr, Nt, and Swiss-Prot databases were each less than 1×10^{-5}, and the thresholds for the Pfam, KOG, GO, and KEGG databases were less than 1×10^{-2}, 1×10^{-3}, 1×10^{-6}, and 1×10^{-10}, respectively.

2.5. Analyses of Differentially Expressed Genes (DEGs)

The clean data were searched against the assembled unigenes, and the read count for each unigene was calculated using the RSEM method [26]. EdgeR software was used to adjust the read counts through one scaling normalized factor. Differential expression analysis was then performed using the DEGSeq R package. The p values were adjusted using the method described by Yoav et al. [27]. The threshold of significant differential expression was q-value < 0.005, $|\log2(\text{fold change})| > 1$.

2.6. KEGG Enrichment Analysis of Differentially Expressed Genes

KEGG and GO (Biological process) enrichment were performed to further understand the function of differentially expressed genes (DEGs). Hypergeometric distribution was used to test the statistical enrichment of DEGs in KEGG pathways.

2.7. Statistical Analysis

The data displayed in the figures were presented as means ± SE. The statistical significance of all data was analyzed using a univariate analysis of variance ($p < 0.05$) (one-way ANOVA; SPSS version 19.0).

3. Results

3.1. The Photosynthesis of B. striata under Drought Stress

After four weeks of exposure to the simulated drought condition, most of the *B. striata* leaves turned yellow and dry, and the roots were short and slim (Figure 1). As illustrated in Figure 2, for the drought-treated plants, a gradual decrease in net photosynthetic rate (*Pn*), stomatal conductance (*Gs*), and transpiration rate (*E*) was recorded throughout the four weeks, except for a sharp decline from days 14 to 21. In addition, the intercellular CO_2 concentration (Ci) gradually increased in the drought treatment but did not significantly change in the CK. Further analysis showed that the value of stomatal limitation, *Ls*, gradually decreased, while the nonstomatal limitation, Ci/*Gs*, sharply increased from days 14 to 21, which suggested that the plant had incurred serious cellular damage after 21 days of drought condition.

Figure 1. Images of *B. striata* under normal (CK) and drought conditions.

3.2. De Novo Assembly, Quality Assessment, and Annotation of Transcriptome

To further investigate the mechanisms that regulate the drought response of *B. striata*, Illumina RNA-Seq sequencing was performed in the present study. Totals of 51,068,510, 47,394,690, 49,121,326, 45,204,114, 56,709,020, and 55,046,454 paired-end reads (150 bp) were obtained in CK1, CK2, CK3, drought1, drought2, and drought3, respectively (Table 1). After removal of the adapter and the low-quality reads, 49,902,994, 46,355,072, 48,119,710, 44,225,174, 55,207,808, and 53,905,350 clean reads were retained in CK1, CK2, CK3, drought1, drought2, and drought3, respectively (Table 1). The value of Q30 in each sample was more than 90%, implying that the clean data were of high quality and could be used for further analysis (Table 1). To assess the reproducibility of the CK and the drought treatments, the Pearson analysis was performed in this study. The results showed that values of R^2 among CK1, CK2, and CK3 ranged from 0.75 to 0.79, while the values among drought1, drought2, and drought3 ranged from 0.74 to 0.80 (Figure 3). These results suggest good reproducibility in the group. Trinity software was used to assemble the clean reads de novo, and 122,160 unigenes with a 1451 bp mean length were obtained. The median length and N50 were 1044 and 2240 bp, respectively. Afterward, the unigenes were aligned to various databases. From a total of 122,160 unigenes, 81,958 (67.09%) unigenes were aligned to the Nr database, 45,174 (36.97%) unigenes were aligned to the NT database, 29,847 (24.43%) unigenes were aligned to the KO database, 57,293 (46.89%) unigenes were aligned to the SwissProt database, 56,063 (45.89%) unigenes were aligned to the PFAM database, 56,063 (45.89%) unigenes were aligned to the PFAM database, and 21,869 (17.90%) unigenes were aligned to the KOG database. In total, 86,618 (70.90%) out of the 122,160 unigenes could be aligned to at least one database (Table 2).

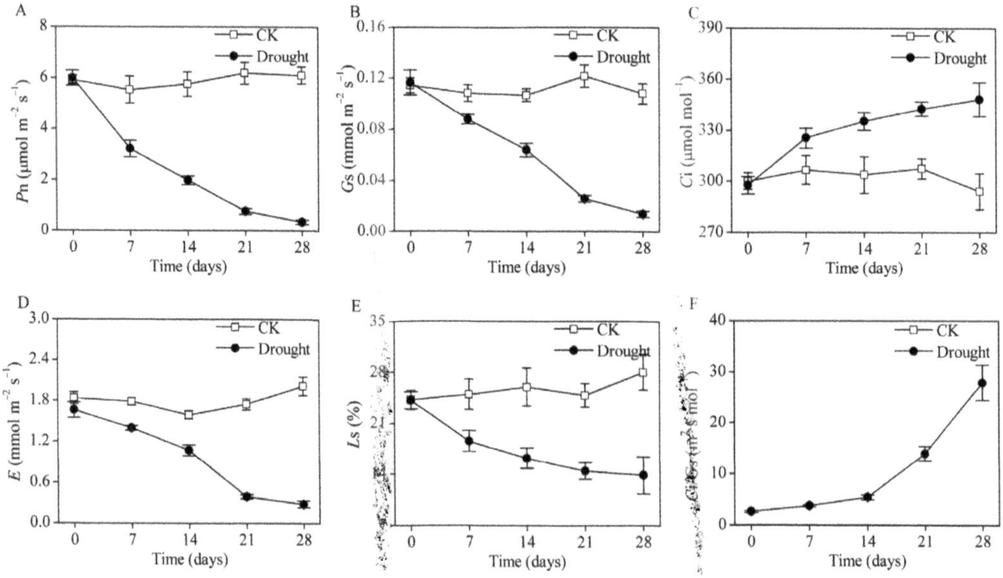

Figure 2. The photosynthetic characteristics of *B. striata* under normal and drought conditions: (**A**) net photosynthetic rate (*Pn*), (**B**) stomatal conductance (*Gs*), (**C**) intercellular CO_2 concentration (Ci), (**D**) evaporation rate (*E*), (**E**) stomatal limitation (*Ls*), and (**F**) ratio of Ci/Gs. Means and standard errors of at least five replicates are shown.

Table 1. Summary of transcriptomes in *B. striata* under drought stress.

Sample	CK1	CK2	CK3	Drought1	Drought2	Drought3
Raw Data	51,068,510	47,394,690	49,121,326	45,204,114	56,709,020	55,046,454
Clean Data	49,902,994	46,355,072	48,119,710	44,225,174	55,207,808	53,905,350
Q30 (%)	92.56	90.14	92.24	92.34	92.40	92.03
GC Content (%)	47.71	47.61	47.65	46.99	47.30	46.97

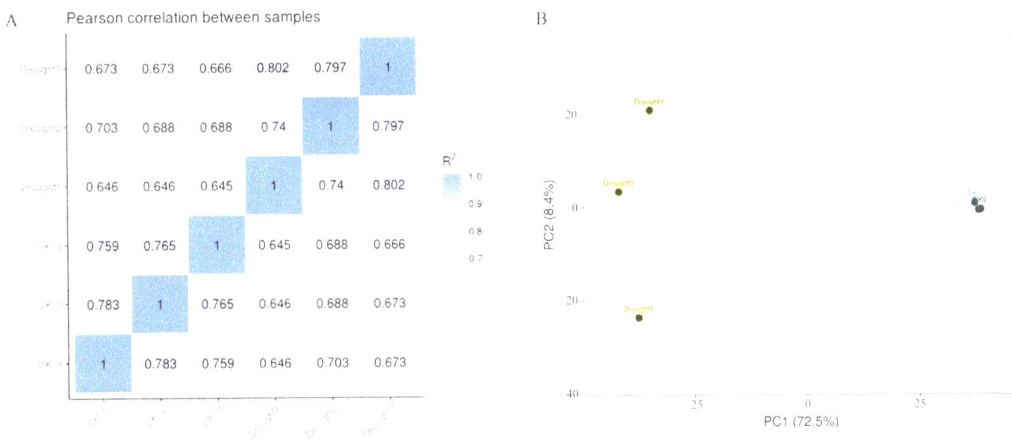

Figure 3. The Pearson correlation coefficient (**A**) and principal component analysis (**B**) of *B. striata* under CK and drought stress.

Table 2. Annotations of unigenes in various databases.

Database	Number of Genes	Percentage (%)
Annotated in NR	81,958	67.09
Annotated in NT	45,174	36.97
Annotated in KO	29,847	24.43
Annotated in SwissProt	57,293	46.89
Annotated in PFAM	56,063	45.89
Annotated in GO	56,063	45.89
Annotated in KOG	21,869	17.9
Annotated in At Least One Database	86,618	70.9

3.3. KEGG and GO Enrichment Analysis of Differentially Expressed Genes (DEGs)

To identify the differentially expressed genes (DEGs) in *B. striata* leaves under drought stress, the expression level of each DEG was calculated. A total of 2398 DEGs were identified, 1271 DEGs were upregulated, and 1127 DEGs were downregulated (Figure 4).

To further explore the underlying mechanism of the *B. striata* response to drought stress, KEGG and GO (Biological process) enrichment analyses were performed. Figure 5A shows that nine KEGG pathways ($p < 0.05$) were significantly enriched in *B. striata* exposed to drought conditions: plant hormone signal transduction (ko04075); phenylpropanoid biosynthesis (ko00940); tropane, piperidine, and pyridine alkaloid biosynthesis (ko00960); phenylalanine metabolism (ko00360); tyrosine metabolism (ko00350); starch and sucrose metabolism (ko00500); cyanoamino acid metabolism (ko00460); diterpenoid biosynthesis (ko00904); and alpha-linolenic acid metabolism (ko00592). Figure 6 presents the expression profiles of the DEGs involved in plant hormone signal transduction. Almost half of the DEGs identified in these pathways were upregulated, and the other half were downregulated, which highlighted the complicated roles of plant hormones in orchestrating the response to drought stress. Most DEGs involved in abscisic-acid signal transduction were also upregulated, such as protein phosphatase 2C (*PP2C*), abscisic acid-insensitive 5-like protein 5 (*ABF*), and serine/threonine protein kinase SAPKs (*SnRK2*).

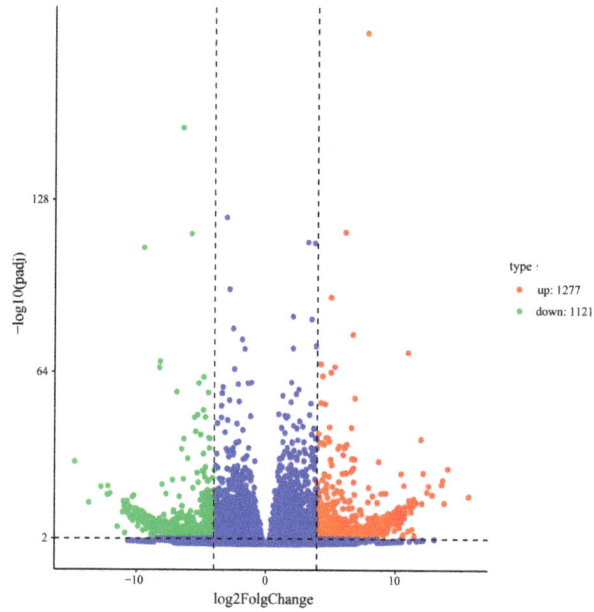

Figure 4. The volcano plot of differentially expressed genes (DEGs) in *B. striata* leaves under CK and drought conditions. Red plots represent upregulated DEGs, green plots represent downregulated DEGs, and blue plots represent the DEGs without significantly change.

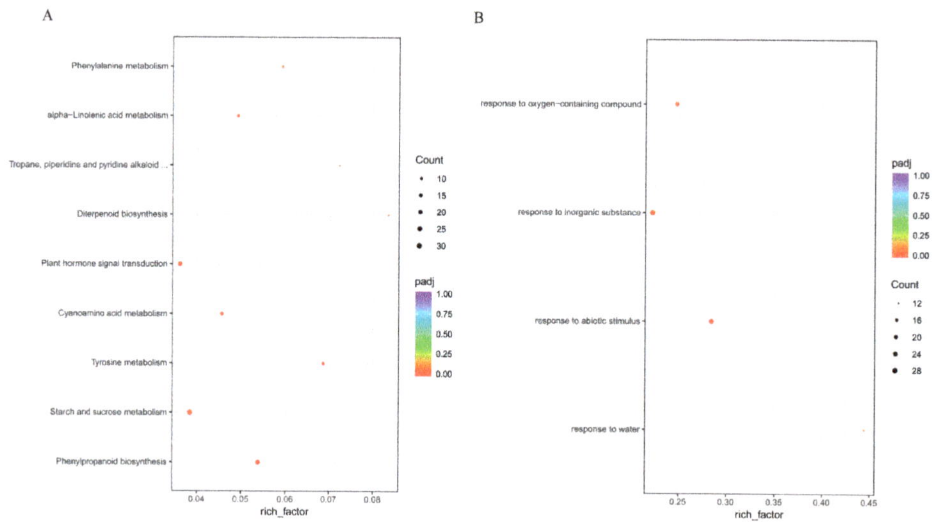

Figure 5. The KEGG and GO enrichment analyses of *B. striata* under CK and drought conditions. (**A**) The results of KEGG enrichment analysis; (**B**) the results of GO (biological process) enrichment analysis.

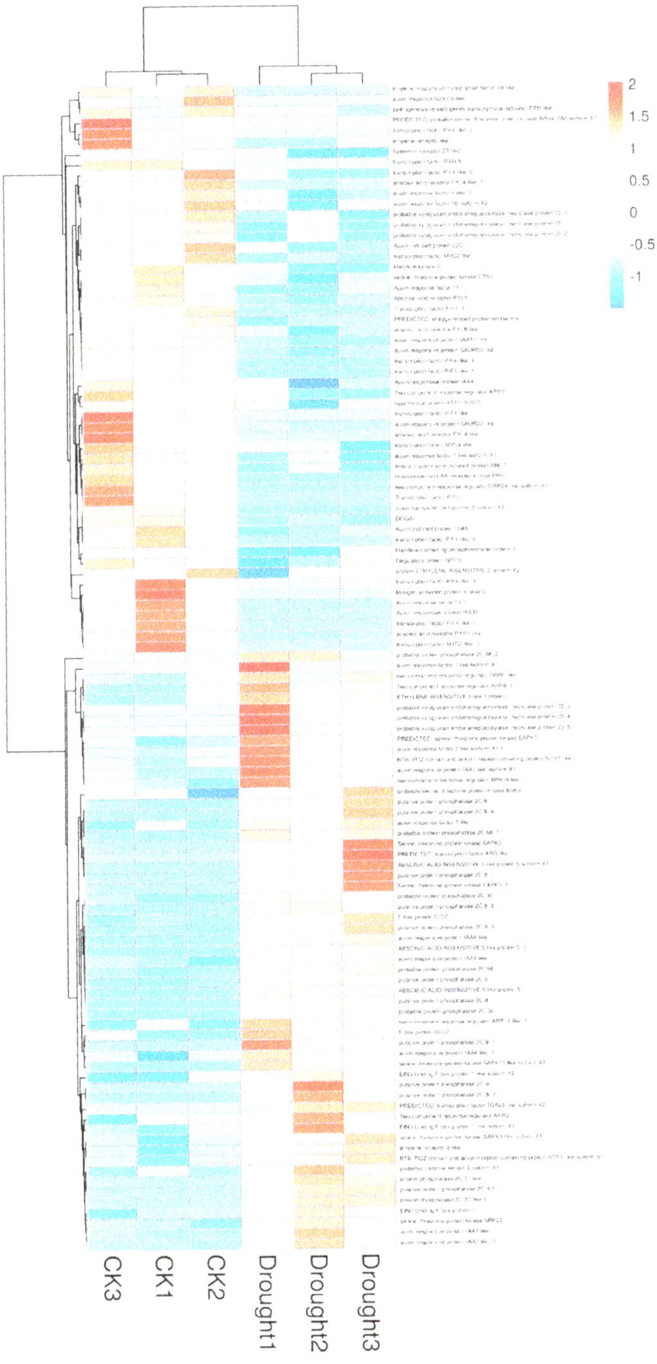

Figure 6. Hierarchical clustering analysis of DEGs involved in plant hormone signal transduction (ko04075) in *B. striata* under CK and drought conditions. Each group has three replicates.

Four GO terms ($p < 0.05$) related to stress are laid out in Figure 5B: response to water (GO:0009415), response to an abiotic stimulus (GO:0009628), response to an inorganic substance (GO:0010035), and response to an oxygen-containing compound (GO:1901700). From there, the expression patterns of DEGs related to these GO terms are presented in Figure 7. Interestingly, we found several DEGs, which encoded dehydrin proteins, were profoundly induced under drought stress (Figure 7).

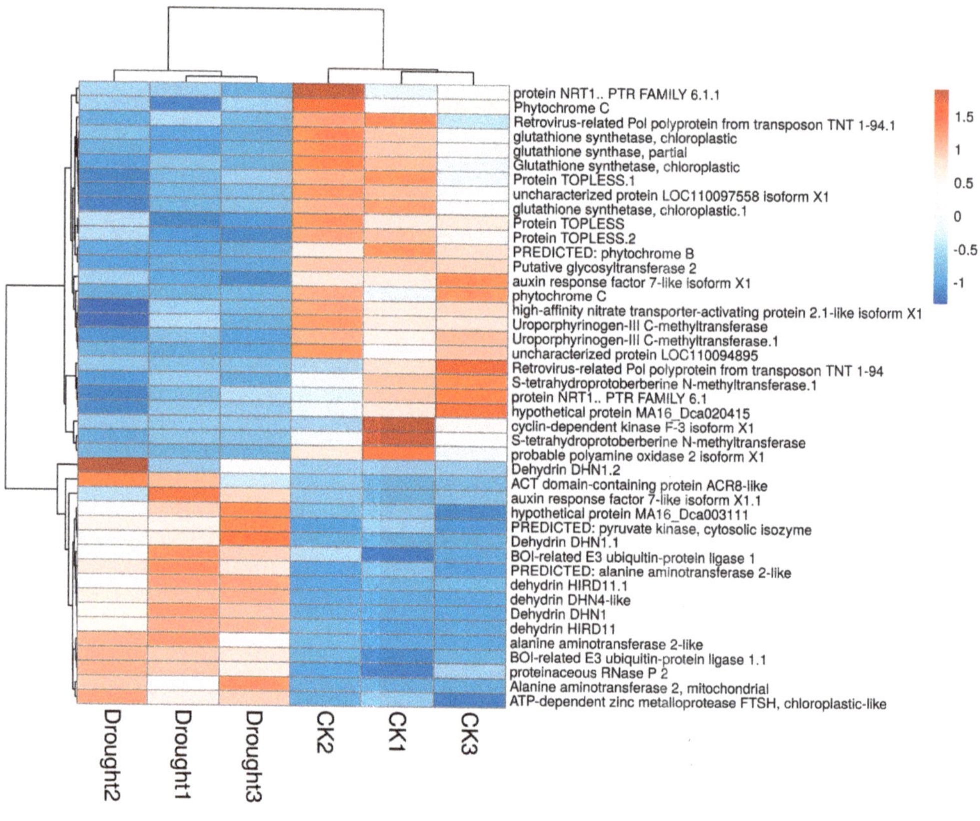

Figure 7. The expression profiles of DEGs involved in four GO terms in *B. striata* under CK and drought conditions, including response to water (GO:0009415), response to an abiotic stimulus (GO:0009628), response to an inorganic substance (GO:0010035), and response to an oxygen-containing compound (GO:1901700). Each group has three replicates.

In addition, detailed information of the DEGs, related to photosynthesis, ABA signal transduction and antioxidant metabolism, are presented in Table 3. Fourteen DEGs were involved in photosynthesis. Four DEGs (Cluster-6724.91253, Cluster-6724.49822, Cluster-6724.49802, and Cluster-6724.50075) were downregulated, and the other ten DEGs were upregulated. Five out of thirty-four DEGs (*PYLs* and *PYR*, which are ABA acceptors) were downregulated in ABA signal transduction, and twenty-nine DEGs (*PP2C*, *ABF*, and *SLAH2*) were upregulated. All DEGs in antioxidant metabolism (*SOD*, *CAT*, *POD*, and *GST*) were upregulated except for Cluster-6724.52388 and Cluster-6724.52387 (*APX*).

Table 3. DEGs related to photosynthesis, ABA signal transduction, and antioxidant metabolism in *B. striata* under drought stress.

Gene ID	NR ID	NR Description	log2FC	*q*-Value
Photosynthesis				
Cluster-6724.91253	YP_009347733.1	Cytochrome b6, chloroplast	−2.2224	0.009795
Cluster-6724.49822	XP_020586841.1	Photosystem I reaction center subunit III, chloroplastic	−1.2025	0.015406
Cluster-6724.49802	XP_020584337.1	Photosystem I reaction center subunit V, chloroplastic	−1.2516	0.000058
Cluster-6724.50075	XP_020592430.1	Photosystem I reaction center subunit psaK, chloroplastic	−1.2483	0.00262
Cluster-6724.44868	XP_020674733.1	psbQ-like protein 3, chloroplastic	7.5757	0.000007
Cluster-6724.68113	PKU73612.1	PsbP domain-containing protein 3, chloroplastic	7.6345	0.026749
Cluster-6724.86842	XP_020701207.1	psbP domain-containing protein 1, chloroplastic isoform X1	1.1728	0.000497
Cluster-6724.55459	PKU68001.1	ATP-dependent zinc metalloprotease FTSH 11, chloroplastic	6.1654	0.012827
Cluster-6724.47025	XP_020682163.1	ATP-dependent zinc metalloprotease FTSH 11, chloroplastic	7.0383	0.018672
Cluster-6724.49015	AVI16663.1	Photosystem I reaction center subunit psaK	3.4885	0.020947
Cluster-6724.74411	XP_020682631.1	Oxygen-evolving enhancer protein 2, chloroplastic-like	1.7217	0.005108
Cluster-2398.0	XP_018676095.1	PREDICTED: photosynthetic NDH subunit of lumenal location 3, chloroplastic-like	5.4806	0.041334
Cluster-6724.11910	XP_020600282.1	Oxygen-evolving enhancer protein 3-2, chloroplastic-like	1.1112	0.030315
Cluster-6724.49786	XP_020672629.1	Photosystem II core complex proteins psbY, chloroplastic isoform X2	1.3762	0.000000
ABA Signal Transduction				
Cluster-6724.40260	XP_020672595.1	Abscisic-acid receptor PYL4-like	−6.9396	0.000001
Cluster-6724.91783	XP_020673631.1	Abscisic-acid receptor PYR1-like	−3.4996	0.000000
Cluster-6724.61155	XP_020587854.1	Abscisic-acid receptor PYL8-like	−1.4327	0.000000
Cluster-6724.38513	PKU64533.1	Abscisic-acid receptor PYL5	−2.9087	0.000000
Cluster-6724.40259	XP_020672595.1	Abscisic-acid receptor PYL4-like	−4.046	0.000000
Cluster-6724.61497	KZV54161.1	Hypothetical protein F511_37072	−1.5431	0.046313
Cluster-6724.3652	PKU68005.1	Putative protein phosphatase 2C 8	2.1736	0.032413
Cluster-6724.16415	PKU68005.1	Putative protein phosphatase 2C 8	3.9737	0.000260
Cluster-6724.91192	XP_020597478.1	Protein phosphatase 2C 37-like	4.6738	0.000000
Cluster-6724.6728	XP_020698862.1	Probable protein phosphatase 2C 68	4.1538	0.000000
Cluster-6724.46469	PKA58960.1	Putative protein phosphatase 2C 8	3.0705	0.000000
Cluster-6724.86228	PKA58960.1	Putative protein phosphatase 2C 8	4.132	0.01917
Cluster-6724.93558	PKU74952.1	Putative protein phosphatase 2C 9	5.5698	0.000110
Cluster-6724.93557	PKU74952.1	Putative protein phosphatase 2C 9	5.4473	0.000036
Cluster-6724.15036	XP_020682015.1	Probable protein phosphatase 2C 30	2.1485	0.000000
Cluster-6724.22798	XP_020698862.1	Probable protein phosphatase 2C 68	8.7029	0.000000
Cluster-6724.22799	XP_020698862.1	Probable protein phosphatase 2C 68	7.5918	0.000025
Cluster-6724.5916	PKU68005.1	Putative protein phosphatase 2C 8	2.9511	0.000000
Cluster-6724.5917	PKU68005.1	Putative protein phosphatase 2C 8	3.25	0.000000
Cluster-6724.5919	PKU68005.1	Putative protein phosphatase 2C 8	3.9231	0.000000
Cluster-6724.90331	XP_020597478.1	Protein phosphatase 2C 37-like	6.6068	0.000000
Cluster-6724.48042	PKU76292.1	Putative protein phosphatase 2C 6	1.9241	0.000000
Cluster-6724.69799	XP_020693557.1	Probable protein phosphatase 2C 50	5.1397	0.000052
Cluster-6724.5918	PKU68005.1	Putative protein phosphatase 2C 8	3.8178	0.000000

Table 3. *Cont.*

Gene ID	NR ID	NR Description	log2FC	q-Value
Cluster-6724.65584	XP_020573821.1	Serine/threonine protein kinase SAPK3-like isoform X1	1.0795	0.041358
Cluster-6724.50267	XP_020705528.1	Serine/threonine protein kinase SAPK10-like isoform X2	2.5198	0.000000
Cluster-6724.56769	XP_015636932.1	PREDICTED: serine/threonine protein kinase SAPK7	1.9686	0.000100
Cluster-6724.18821	PKU80471.1	Serine/threonine protein kinase SAPK3	5.6694	0.004812
Cluster-6724.60505	API65110.1	Serine/threonine protein kinase SRK2E	1.0474	0.000001
Cluster-6724.52592	PKU80471.1	Serine/threonine protein kinase SAPK3	6.8354	0.000240
Cluster-6724.41934	PKU79905.1	ABSCISIC ACID-INSENSITIVE 5-like protein 5	1.2712	0.000000
Cluster-6724.51613	XP_020694098.1	ABSCISIC ACID-INSENSITIVE 5-like protein 5 isoform X1	3.9425	0.008781
Cluster-6724.93796	PKU83951.1	ABSCISIC ACID-INSENSITIVE 5-like protein 5	3.2259	0.000000
Cluster-7290.0	PKU75117.1	S-type anion channel SLAH2	4.146	0.000001
Antioxidant Metabolism				
Cluster-6724.52388	ACN25039.1	Ascorbate peroxidase	−4.3302	0.000005
Cluster-6724.52387	ACN25039.1	Ascorbate peroxidase	−9.2054	0.000000
Cluster-6724.51106	XP_020590426.1	Superoxide dismutase [Cu-Zn] 4A	7.6369	0.000002
Cluster-6724.49968	XP_020702876.1	Catalase isozyme A	7.8841	0.000000
Cluster-4051.0	XP_020585759.1	Peroxidase P7-like isoform X1	4.5187	0.000000
Cluster-6724.2897	PKU65314.1	Peroxidase 42	4.6461	0.000000
Cluster-6724.98401	PKU59654.1	Cationic peroxidase 1	7.797	0.000086
Cluster-6724.78777	XP_020679253.1	Probable glutathione S-transferase parA	9.8673	0.000006
Cluster-6724.95492	PKU87189.1	Putative glutathione S-transferase parA	6.9382	0.003953
Cluster-6724.45643	PKU87189.1	Putative glutathione S-transferase parA	4.1034	0.002935
Cluster-15974.0	XP_020573757.1	Glutathione S-transferase F8, chloroplastic-like	4.5425	0.000004

4. Discussion

4.1. ABA Signal Transduction in B. striata Leaves under Drought Stress

Phytohormones regulate various biological processes to control plant growth and stress responses [28]. To survive under drought stress, plants have evolved extensive phytohormone signaling pathways in response to drought stress, such as those of auxin (IAA), cytokinin (CK), abscisic acid (ABA), jasmonic acid (JA), salicylic acid (SA), ethylene (ET), and gibberellin (GA) [29,30]. The present study identified a series of DEGs that participate in plant hormone signal transduction (ko04075) (Figures 5A and 6; detailed information for ko04075 is shown in Supplementary Table S2). Among these phytohormones, ABA was regarded as the major stress-responsive hormone under drought stress [30]. Under normal conditions, PP2C would inhibit the activity of SnRK2 protein via dephosphorylating [30]. When a plant is exposed to drought conditions, its cellular ABA concentration increases, which binds PYL/PYR/RCARs proteins. This ABA-PYL/PYR/RCARs complex would then inhibit the activity of PP2C, which would lead to the activation of SnRK2 [31,32]. The activated SnRK2 would phosphorylate downstream genes and trigger the ABA-induced response [30]. A recent study reported that PYL-PP2C-SnRK2s, which function as the ABA core signal pathway, also exist in *B. striata* [21]. Our results also showed that many DEGs engaged in ABA signal transduction (*PP2C, ABF, SnRK2*), and most of them were upregulated except for *PYL/PYR/RCARs* (Figure 6). The *SnRK2* gene is suspected to be positively related to stomatal closure, since a *srk2e* mutation in *Arabidopsis* resulted in a wilty mutant caused by a loss of stomatal closure under drought stress [33]. Upregulated *SnRK2* could activate several cation or anion channels, such as S-type anion channels (SLAHs), to force the stomata to close [34]. Based on these results, the upregulations of *SnRK2* and *SLAH2* (Table 3) in *B. striata* under drought stress suggest that drought stress might result in promoting stomatal closure to prevent water loss from leaves via controlling ion efflux in guard cells. Mori et al. (2006) also found that ABA was initially synthesized in the roots and subsequently migrated to the leaves, where it would shut the stomata and reduce plant photosynthesis [35]. Consistently with previous studies, our results found that most of the DEGs (*PP2C*, *SnRK2*, and *SLAH2*) that participated in core ABA signaling pathway PYL-PP2C-SnRK2 were upregulated (Table 3), and the values of the net photosynthetic rate (*Pn*) and stomatal conductance (*Gs*) were decreased under drought stress (Figure 2). The same results were also reported by Liu et al. [21]; stomatal closure was gradually decreased in *B. striata* leaves with lower soil water content. The decline of *Pn* could have been caused by stomatal limitation (*Ls*) and nonstomatal limitation (Ci/*Gs*) [36]. Our results found a decrease in *Ls* but an increase in Ci/*Gs* in *B. striata* leaves under drought stress (Figure 2). According to the results presented above, we concluded that the decline of photosynthesis in *B. striata* leaves under drought stress is mainly caused by nonstomatal limitation factors, which are mediated with the ABA signaling pathway.

Moreover, SnRK2 could activate downstream genes such as *ABF* [37]. Overexpression of *ABF* in *Arabidopsis* resulted in ABA hypersensitivity and high drought tolerance [38]. Meanwhile, upregulated *SnRK2* could activate several cation or anion channels, such as S-type anion channel 3 (SLAH3), to force stomata to close [34]. In the present study, the upregulation of *ABFs* indicated that the high expression level of *ABFs* in *B. striata* leaves could enhance the plant's drought tolerance under drought stress.

4.2. *Effect of Drought Stress on DEGs Involved in Photosynthesis*

As we mentioned above, nonstomatal limitation factors were the main reason for photosynthesis reduction. In other words, there was metabolic damage in the photosynthetic process, such as downregulation of gene expression of some photosynthesis-related proteins. Our transcriptional analyses found that four DEGs that participated in photosynthesis were downregulated. Cytochrome b6 was reported to function in regulating electron transfer between photosystem II and photosystem I [39]. The stability of *petB* transcripts could control cytochrome b6 levels. The transcriptional level of Cytochrome b6 (*petB*, Cluster-6724.91253) was decreased in our study, suggesting that the level of cytochrome b6 is reduced under drought stress and leads to an inhibition of electron transfer between photosystem II and photosystem I, which would result in low photosynthesis in *B. striata*. Photosystem I reaction center subunit III (*psaF*, Cluster-6724.49822) is a plastocyanin-docking protein participating in regulating efficiency of electron transfer from plastocyanin to P700 [40]. The lack of *psaF* results in an inability of energy transfer from light-harvesting complex I-730 to the P700 reaction center [40]. Photosystem I reaction center subunit IV (*psaE*, Cluster-6724.49802) participates in docking of ferredoxin to PSI and interaction with ferredoxin-NADP oxidoreductase [41]. Absence of *psaE* leads to low O_2 production and serious damage in PSII under photoinhibition conditions [41]. In other words, low transcriptional levels of *psaE* are adverse for plant photosynthesis. The downregulations of *psaE* and *psaF* in the present study imply that drought stress inhibits energy transfer and PSI assembling via regulating the expressions of *psaE* and *psaF* and subsequently affecting the photosynthesis of *B. striata*. As Krieger-Liszkav (2020) pointed out, high O_2 production protects PSII against photoinhibition [41]. Our transcriptomic data also found several DEGs related to O_2 production in *B. striata* leaves under drought stress. For example, psbP-domain-containing proteins (Cluster-6724.68113, Cluster-6724.86842, and Cluster-6724.74411) and psbQ proteins (Cluster-6724.44868 and Cluster-6724.11910) were reported to play a functional role in optimization of photosynthetic oxygen evolution [42,43]. Interestingly, all of these DEGs were upregulated, which will lead to higher efficiency of O_2 production, suggesting that *B. striata* might enhance O_2 production to protect its photosystem under drought stress. In addition, ATP-dependent zinc metalloprotease FTSH 11 (Cluster-6724.55459, Cluster-6724.47025) was found to be upregulated in *B. striata* leaves under drought stress (Table 3). In *Arabidopsis*, FtsH6 was found to participate in degradation of light-harvesting complex II during high-light acclimation [44].

Upregulation of FtsH11, a homologous protein, might have the same function in regulating degradation of light-harvesting complex II, which would lead to low efficiency of light-energy absorption. These changes in gene expression may be a self-protection mechanism for plants under drought stress.

4.3. Drought-Induced Gene Expression of Stress-Response Protein

When a plant is exposed to drought stress, many stress-responsive genes are induced to protect it from drought-induced damage [10]. Plants generally overaccumulate ROS in their tissues after any stressful insult, both biotic and abiotic. An antioxidant battery is the first line of defense against oxidative damage in plant cells [10]. The antioxidants commonly utilized by plants include catalase (CAT), superoxide dismutase (SOD), peroxidase (POX), ascorbate peroxidase (APX), glutathione sulfo-transferase (GST), and others [9]. As displayed in Table 3, almost all DEGs related to antioxidant metabolism were significantly upregulated except for *APX*. Overproduction of O^{2-} is the first step undertaken by a plant under drought stress. SOD catalyzes O^{2-} into the significantly less toxic H_2O_2 [12]. As more and more H_2O_2 is produced in a cell, it becomes reduced into H_2O with the enzymatic antioxidant CAT through electron transport, photorespiratory oxidation, and oxidation of fatty acids [7]. Alternatively, H_2O_2 can also be converted into H_2O with APX via the AsA-GSH cycle [7]. Under drought conditions, however, the expression level of *APX* was decreased, whereas the expression levels of *SOD* and *CAT* were increased, implying that ROS produced in drought-stressed *B. striata* leaves is mainly scavenged by SOD and CAT and not via the APX-mediated AsA-GSH cycle. In addition, POX was reported to minimize drought-induced cellular damage due to its ability to lignify and crosslink structural proteins in cell walls [11]. These results indicate that *B. striata* leaves reduce drought-induced oxidative damage via activating expressions of *SOD*, *CAT*, and *POX* to scavenge ROS.

Moreover, GO enrichment analysis identified six dehydrin (*DHN*) genes, which were highly expressed under drought stress (Figures 5 and 7). DHNs were initially recognized as "dehydration-induced proteins" in response to desiccation [45]. An increasing body of evidence suggests that DHN proteins impart drought stress tolerance through activating various biological processes, such as photosynthesis, ROS scavenging, accumulation of compatible solutes, and others [46]. Overexpression of *DHNs* could significantly enhance drought tolerance in plants. For example, overexpressions of two dehydrins, *Y2SK2* and *SK3*, in *Arabidopsis thaliana* resulted in higher tolerance of salt, osmotic cold, and drought stress, with higher antioxidant activity and photosynthesis [47]. In addition, DHN1 was reported to maintain high chlorophyll content and water fresh/dry weight but low H_2O_2 concentration, resulting from enhanced ROS scavenging [48]. Similar to these findings, we found that *DHNs* were significantly engaged in *B. striata* leaves under drought stress, implying that a high expression level of *DHNs* could enhance drought tolerance through enhancing ROS scavenging, which was consistent with our results of high expression levels of *SOD*, *CAT*, and *POX*.

5. Conclusions

B. striata is a valuable Chinese herbal medicinal plant. During the planting process, drought is one of the most serious threats to its growth and development. In the present study, the response of *B. striata* under drought stress was investigated at the physiological and transcriptional levels. Photosynthetic results indicated that the decline of photosynthesis in *B. striata* leaves was mainly caused by nonstomatal limitation factors. Transcriptomic analysis showed that DEGs involved in photosynthesis processes, such as electron transfer (cytochrome b6) and light-energy harvesting and transfer (photosystem I reaction center subunit III and photosystem I reaction center subunit IV), might lead to reduction in *B. striata* photosynthesis, while DEGs related to O_2 production (*psbP*, *psbQ*) and light-energy absorption (ATP-dependent zinc metalloprotease *FTSH 11*) are activated to protect the plant from drought stress. Moreover, the DEGs involved in the ABA signaling pathway were the most upregulated. Upregulations of *PP2C*, *SnRK2*, and *SLAH2* might lead to stomatal closure, which is one of the reasons for photosynthesis reduction. In response to drought stress, the *B. striata* leaves recruited SOD, CAT, and POX, which enhanced the ability of ROS scavenging. High expressions of ABF and DHNs might result in high drought tolerance in *B. striata*. According to these results, a better understanding of *B. striata* in response to drought stress was presented in our study.

Supplementary Materials: The following supporting information can be downloaded at: https://www.mdpi.com/article/10.3390/horticulturae9030307/s1, Table S1: The detailed information of KEGG and GO enrichment analyses, Table S2: The information of DEGs involved in plant hormone signal transduction.

Author Contributions: Conceptualization, Z.S. and M.W.; methodology, H.L. and L.Y.; software, Z.S. and H.L.; validation, X.H., K.C. and Z.S.; formal analysis, M.W.; investigation, H.L. and K.C.; resources, M.W. and H.L.; data curation, M.W. and Z.S.; writing—original draft preparation, H.L. and K.C.; writing—review and editing, K.C., X.H., Z.S. and M.W.; visualization, H.L. and X.H.; supervision, Z.S. and M.W.; project administration, H.L. and M.W.; funding acquisition, M.W. All authors have read and agreed to the published version of the manuscript.

Funding: This research was funded by the National Key Research and Development Program of China (Grant No. 2021YFD1601002), the Guizhou Provincial Department of Science and Technology (Grant No. QKHFQ2020-4008, QKHZC2022-ZD023), the Modern Industrial Technology System of Chinese Medicinal Materials in Guizhou Province (No. GZCYTX2021-0202), the Natural Science Foundation of China (Grant No. 31401362), and the Major Science and Technology Program of Xiamen, China (Grant No. 3502Z20211004).

Data Availability Statement: The raw sequence data reported in this paper have been deposited in the Genome Sequence Archive (Genomics, Proteomics & Bioinformatics 2021) in the National Genomics Data Center (Nucleic Acids Res 2022), China National Center for Bioinformation/Beijing Institute of Genomics, Chinese Academy of Sciences (GSA: CRA008921), which is publicly accessible at https://ngdc.cncb.ac.cn/gsa.

Conflicts of Interest: The authors declare no conflict of interest.

References

1. Gupta, A.; Rico-Medina, A.; Caño-Delgado, A.I. The physiology of plant responses to drought. *Science* **2020**, *368*, 266–269. [CrossRef] [PubMed]
2. Hussain, M.; Malik, M.A.; Farooq, M.; Ashraf, M.Y.; Cheema, M.A. Improving Drought Tolerance by Exogenous Application of Glycinebetaine and Salicylic Acid in Sunflower. *J. Agron. Crop. Sci.* **2008**, *194*, 193–199. [CrossRef]
3. Cattivelli, L.; Rizza, F.; Badeck, F.-W.; Mazzucotelli, E.; Mastrangelo, A.M.; Francia, E.; Marè, C.; Tondelli, A.; Stanca, A.M. Drought tolerance improvement in crop plants: An integrated view from breeding to genomics. *Field Crop. Res.* **2008**, *105*, 1–14. [CrossRef]
4. Mckay, J.K.; Richards, J.H.; Mitchell-Olds, T. Genetics of drought adaptation in *Arabidopsis thaliana*: I. Pleiotropy contributes to genetic correlations among ecological traits. *Mol. Ecol.* **2003**, *12*, 1137–1151. [CrossRef]
5. Prasad, P.V.V.; Staggenborg, S.; Ristic, Z. Impacts of drought and/or heat stress on physiological, developmental, growth, and yield processes of crop plants. *Adv. Agric. Syst. Model. Ser.* **2008**, *1*, 301–355.
6. Shao, H.-B.; Chu, L.-Y.; Jaleel, C.A.; Zhao, C.-X. Water-deficit stress-induced anatomical changes in higher plants. *Comptes Rendus Biol.* **2008**, *331*, 215–225. [CrossRef]
7. Laxa, M.; Liebthal, M.; Telman, W.; Chibani, K.; Dietz, K.-J. The Role of the Plant Antioxidant System in Drought Tolerance. *Antioxidants* **2019**, *8*, 94. [CrossRef]
8. Corpas, F.J.; González-Gordo, S.; Palma, J.M. Plant Peroxisomes: A Factory of Reactive Species. *Front. Plant Sci.* **2020**, *11*, 853. [CrossRef]
9. Lukatkin, A.S.; Anjum, N.A. Control of cucumber (*Cucumis sativus* L.) tolerance to chilling stressâ€"evaluating the role of ascorbic acid and glutathione. *Front. Environ. Sci.* **2014**, *2*, 62. [CrossRef]
10. Mukarram, M.; Choudhary, S.; Kurjak, D.; Petek, A.; Khan, M.M.A. Drought: Sensing, signalling, effects and tolerance in higher plants. *Physiol. Plant.* **2021**, *172*, 1291–1300. [CrossRef]
11. Quan, L.-J.; Zhang, B.; Shi, W.-W.; Li, H.-Y. Hydrogen Peroxide in Plants: A Versatile Molecule of the Reactive Oxygen Species Network. *J. Integr. Plant Biol.* **2008**, *50*, 2–18. [CrossRef]
12. Zandalinas, S.I.; Balfagón, D.; Arbona, V.; Gómez-Cadenas, A. Modulation of Antioxidant Defense System Is Associated with Combined Drought and Heat Stress Tolerance in Citrus. *Front. Plant Sci.* **2017**, *8*, 953. [CrossRef]

13. Chen, Z.; Cheng, L.; He, Y.; Wei, X. Extraction, characterization, utilization as wound dressing and drug delivery of *Bletilla striata* polysaccharide: A review. *Int. J. Biol. Macromol.* **2018**, *120*, 2076–2085. [CrossRef] [PubMed]
14. Zhang, C.; Gao, F.; Gan, S.; He, Y.; Chen, Z.; Liu, X.; Fu, C.; Qu, Y.; Zhang, J. Chemical characterization and gastroprotective effect of an isolated polysaccharide fraction from *Bletilla striata* against ethanol-induced acute gastric ulcer. *Food Chem. Toxicol.* **2019**, *131*, 110539. [CrossRef] [PubMed]
15. Liu, H.; Li, L.; Li, C.; Huang, C.; ShangGuan, Y.; Chen, R.; Xiao, S.; Wen, W.; Xu, D. Identification and bioinformatic analysis of *Aux/IAA* family based on transcriptome data of *Bletilla striata*. *Bioengineered* **2019**, *10*, 668–678. [CrossRef] [PubMed]
16. Niu, J.; Zhao, G.; Mi, Z.; Chen, L.; Liu, S.; Wang, S.; Wang, D.; Wang, Z. De novo sequencing of *Bletilla striata* (Orchidaceae) transcriptome and identification of genes involved in polysaccharide biosynthesis. *Genet. Mol. Biol.* **2020**, *43*, e20190417. [CrossRef] [PubMed]
17. Jiang, L.; Lin, M.; Wang, H.; Song, H.; Zhang, L.; Huang, Q.; Chen, R.; Song, C.; Li, G.; Cao, Y. Haplotype-resolved genome assembly of *Bletilla striata* (Thunb.) Reichb.f. to elucidate medicinal value. *Plant J.* **2022**, *111*, 1340–1353. [CrossRef]
18. Ding, L.; Shan, X.; Zhao, X.; Zha, H.; Chen, X.; Wang, J.; Cai, C.; Wang, X.; Li, G.; Hao, J.; et al. Spongy bilayer dressing composed of chitosan–Ag nanoparticles and chitosan–*Bletilla striata* polysaccharide for wound healing applications. *Carbohydr. Polym.* **2017**, *157*, 1538–1547. [CrossRef]
19. Wang, C.X.; Tian, M.; Li, Q.J.; Liu, F. Floral syndrome and breeding system of *Bletilla striata*. *Acta Hortic. Sin.* **2012**, *39*, 1159–1166.
20. Zhang, M.; Shao, Q.; Xu, E.; Wang, Z.; Wang, Z.; Yin, L. *Bletilla striata*: A review of seedling propagation and cultivation modes. *Physiol. Mol. Biol. Plants* **2019**, *25*, 601–609. [CrossRef]
21. Liu, S.; Lu, C.; Jiang, G.; Zhou, R.; Chang, Y.; Wang, S.; Wang, D.; Niu, J.; Wang, Z. Comprehensive functional analysis of the PYL-PP2C-SnRK2s family in *Bletilla striata* reveals that BsPP2C22 and BsPP2C38 interact with BsPYLs and BsSnRK2s in response to multiple abiotic stresses. *Front. Plant Sci.* **2022**, *13*, 963069. [CrossRef] [PubMed]
22. Gao, Y.; Cai, C.; Yang, Q.; Quan, W.; Li, C.; Wu, Y. Response of *Bletilla striata* to Drought: Effects on Biochemical and Physiological Parameter Also with Electric Measurements. *Plants* **2022**, *11*, 2313. [CrossRef]
23. Larocque, G.R. Coupling a detailed photosynthetic model with foliage distribution and light attenuation functions to compute daily gross photosynthesis in sugar maple (*Acer saccharum* Marsh.) stands. *Ecol. Model.* **2002**, *148*, 213–232. [CrossRef]
24. Ramanjulu, S.; Sreenivasulu, N.; Sudhakar, C. Effect of Water Stress on Photosynthesis in Two Mulberry Genotypes with Different drought Tolerance. *Photosynthetica* **1998**, *35*, 279–283. [CrossRef]
25. Grabherr, M.G.; Haas, B.J.; Yassour, M.; Levin, J.Z.; Thompson, D.A.; Amit, I.; Adiconis, X.; Fan, L.; Raychowdhury, R.; Zeng, Q.D.; et al. Full-length transcriptome assembly from RNA-Seq data without a reference genome. *Nat. Biotechnol.* **2011**, *29*, 644–652. [CrossRef] [PubMed]
26. Li, B.; Dewey, C.N. RSEM: Accurate transcript quantification from RNA-Seq data with or without a reference genome. *BMC Bioinform.* **2011**, *12*, 323. [CrossRef] [PubMed]
27. Benjamini, Y.; Drai, D.; Elmer, G.; Kafkafi, N.; Golani, I. Controlling the false discovery rate in behavior genetics research. *Behav. Brain Res.* **2001**, *125*, 279–284. [CrossRef]
28. Weyers, J.D.B.; Paterson, N.W. Plant hormones and the control of physiological processes. *New Phytol.* **2001**, *152*, 375–407. [CrossRef]
29. Kazan, K. Diverse roles of jasmonates and ethylene in abiotic stress tolerance. *Trends Plant Sci.* **2015**, *20*, 219–229. [CrossRef]
30. Ullah, A.; Manghwar, H.; Shaban, M.; Khan, A.H.; Akbar, A.; Ali, U.; Ali, E.; Fahad, S. Phytohormones enhanced drought tolerance in plants: A coping strategy. *Environ. Sci. Pollut. Res.* **2018**, *25*, 33103–33118. [CrossRef]
31. Danquah, A.; de Zelicourt, A.; Colcombet, J.; Hirt, H. The role of ABA and MAPK signaling pathways in plant abiotic stress responses. *Biotechnol. Adv.* **2014**, *32*, 40–52. [CrossRef] [PubMed]
32. Ullah, A.; Sun, H.; Yang, X.; Zhang, X. Drought coping strategies in cotton: Increased crop per drop. *Plant Biotechnol. J.* **2017**, *15*, 271–284. [CrossRef] [PubMed]
33. Yoshida, R.; Hobo, T.; Ichimura, K.; Mizoguchi, T.; Takahashi, F.; Aronso, J.; Ecker, J.; Shinozaki, K. ABA-Activated SnRK2 Protein Kinase is Required for Dehydration Stress Signaling in *Arabidopsis*. *Plant Cell Physiol.* **2002**, *43*, 1473–1483. [CrossRef]
34. Bharath, P.; Gahir, S.; Raghavendra, A.S. Abscisic Acid-Induced Stomatal Closure: An Important Component of Plant Defense Against Abiotic and Biotic Stress. *Front. Plant Sci.* **2021**, *12*, 615114. [CrossRef]
35. Mori, I.C.; Murata, Y.; Yang, Y.; Munemasa, S.; Wang, Y.-F.; Andreoli, S.; Tiriac, H.; Alonso, J.M.; Harper, J.F.; Ecker, J.R.; et al. CDPKs CPK6 and CPK3 Function in ABA Regulation of Guard Cell S-Type Anion- and Ca2+- Permeable Channels and Stomatal Closure. *PLoS Biol.* **2006**, *4*, 1749–1762. [CrossRef] [PubMed]
36. Flexas, J.; Medrano, H. Drought-inhibition of photosynthesis in C3 plants: Stomatal and non-stomatal limitations revisited. *Ann. Bot.* **2002**, *89*, 183–189. [CrossRef] [PubMed]
37. de Ollas, C.; Dodd, I.C. Physiological impacts of ABA–JA interactions under water-limitation. *Plant Mol. Biol.* **2016**, *91*, 641–650. [CrossRef] [PubMed]
38. Fujita, Y.; Fujita, M.; Satoh, R.; Maruyama, K.; Parvez, M.M.; Seki, M.; Hiratsu, K.; Ohme-Takagi, M.; Shinozaki, K.; Yamaguchi-Shinozaki, K. AREB1 Is a Transcription Activator of Novel ABRE-Dependent ABA Signaling That Enhances Drought Stress Tolerance in *Arabidopsis*. *Plant Cell* **2005**, *17*, 3470–3488. [CrossRef]

39. Stoppel, R.; Lezhneva, L.; Schwenkert, S.; Torabi, S.; Felder, S.; Meierhoff, K.; Westhoff, P.; Meurer, J. Recruitment of a Ribosomal Release Factor for Light- and Stress-Dependent Regulation of *petB* Transcript Stability in *Arabidopsis* Chloroplasts. *Plant Cell* **2011**, *23*, 2680–2695. [CrossRef]
40. Haldrup, A.; Simpson, D.J.; Scheller, H.V. Down-regulation of the PSI-F Subunit of Photosystem I (PSI) in *Arabidopsis thaliana*—The PSI-F subunit is essential for photoautotrophic growth and contributes the antenna function. *J. Biol. Chem.* **2000**, *275*, 31211–31218. [CrossRef]
41. Krieger-Liszkay, A.; Shimakawa, G.; Sétif, P. Role of the two PsaE isoforms on O_2 reduction at photosystem I in *Arabidopsis thaliana*. *Biochim. Biophys. Acta (BBA)—Bioenerg.* **2019**, *1861*, 148089. [CrossRef] [PubMed]
42. Bricker, T.M.; Roose, J.L.; Zhang, P.; Frankel, L.K. The PsbP family of proteins. *Photosynth. Res.* **2013**, *116*, 235–250. [CrossRef] [PubMed]
43. Kakiuchi, S.; Uno, C.; Ido, K.; Nishimura, T.; Noguchi, T.; Ifuku, K.; Sato, F. The PsbQ protein stabilizes the functional binding of the PsbP protein to photosystem II in higher plants. *Biochim. Biophys. Acta (BBA)—Bioenerg.* **2012**, *1817*, 1346–1351. [CrossRef] [PubMed]
44. Żelisko, A.; García-Lorenzo, M.; Jackowski, G.; Jansson, S.; Funk, C. AtFtsH6 is involved in the degradation of the light-harvesting complex II during high-light acclimation and senescence. *Proc. Natl. Acad. Sci. USA* **2005**, *102*, 13699–13704. [CrossRef]
45. Close, T.J.; Kortt, A.A.; Chandler, P.M. A cDNA-based comparison of dehydration-induced proteins (dehydrins) in barley and corn. *Plant Mol. Biol.* **1989**, *13*, 95–108. [CrossRef]
46. Riyazuddin, R.; Nisha, N.; Singh, K.; Verma, R.; Gupta, R. Involvement of dehydrin proteins in mitigating the negative effects of drought stress in plants. *Plant Cell Rep.* **2021**, *41*, 519–533. [CrossRef]
47. Yang, Z.; Sheng, J.; Lv, K.; Ren, L.; Zhang, D. Y2SK2 and SK3 type dehydrins from *Agapanthus praecox* can improve plant stress tolerance and act as multifunctional protectants. *Plant Sci.* **2019**, *284*, 143–160. [CrossRef]
48. Verma, G.; Dhar, Y.V.; Srivastava, D.; Kidwai, M.; Chauhan, P.S.; Bag, S.K.; Asif, M.H.; Chakrabarty, D. Genome-wide analysis of rice dehydrin gene family: Its evolutionary conservedness and expression pattern in response to PEG induced degydration stress. *PLoS ONE* **2017**, *12*, e0176399. [CrossRef]

horticulturae

MDPI

Article

Stepwise Optimization of the RT-qPCR Protocol and the Evaluation of Housekeeping Genes in Pears (*Pyrus bretschneideri*) under Various Hormone Treatments and Stresses

Peng Zhou [1], Linlin Huang [1], Yingtao Wang [2], Xiao Li [2], Xinxin Feng [1] and Liulin Li [1,*]

1 College of Horticulture, Shanxi Agricultural University, Taigu 030801, China
2 Shijiazhuang Institute of Fruit Trees, Hebei Academy of Agriculture and Forestry, Shijiazhuang 050061, China
* Correspondence: tgliulin@163.com

Abstract: Real-time quantitative PCR (RT-qPCR) has become a widely used method for exploring plant gene expression level. The method requires using some stably expressed genes as a reference to accurately normalize the RT-qPCR data. However, under various stresses and hormone treatments, the levels of most reference genes vary. Environmental variations also influence their expression levels. The lack of validated, stably expressed reference genes can mislead the study of gene function in pears. "Huangguan" pears have recently become the focus of research on stress resistance mechanisms, such as high resistance. Therefore, the aim of this study was to select the optimal reference genes in Huangguan pears, and we analyzed the expression of the genes *EF1α, ACT, SKD1, YLS8, UBQ, GAPDH, TUB*, and *WDP* in a series of pear leaf sets under various stresses and hormone treatments. Using different statistical algorithms, we found that under various treatments, the *WDP* gene had more stable expression, *ACT* was the most stable under MeJA treatment, *YLS8* was the most valuable reference gene under ABA hormone and heat stress conditions, and *GAPDH* showed worst results compared to other housekeeping genes, except under heat stress. These results will supply valuable and updated information for the selection of housekeeping genes in pears under biotic and abiotic stresses in the future.

Keywords: housekeeping gene; RT-qPCR; hormone; biotic and abiotic stress; pear

1. Introduction

Pears are one of the most nutritious fruits in the world, providing fiber which is beneficial to human health [1]. Approximately 39 billion tons of pears are delivered across the world every year. Huangguan pear (*Pyrus bretschneideri* Rehd. cv. Huangguan) is a highly resistant new cultivar widely planted in northern China [2]. In nature, biotic and abiotic environmental stresses are critical factors responsible for decreasing fruit quality and yield and increasing tree losses and production costs. Attacks by pests and pathogen with a wide range of infection strategies have caused serious diseases in pears. To make matters worse, their virulence mechanisms constantly evolve to enable them to adapt to their hosts. This makes it challenging to develop effective and long-term control measures [3]. Abiotic stresses, including drought [4], cold [5], and salt [6], negatively affect, for example, the growth, development, and productivity of pears, while biotic stresses, such as pathogens [7] and insects [8], often lead to diseases and damage the crops. Pear scab disease, as a biotic stress, is a devastating disease caused by *Venturia nashicola* in Asian countries [9]. It often leads to enormous economic losses for the pear industry [10]. Phytohormones are key regulators of every aspect of a plant's life cycle [11], including growth, defense, and death. Hormones such as abscisic acid (ABA) [12], ethylene (ETH) [13,14], salicylic acid (SA) [15], and methyl jasmonate (MeJA) [16] are usually associated with plants experiencing biotic and abiotic stresses. Hence, hormones are also used as experimental treatments in the research on defense-related gene expression to explore the mechanisms of metabolic regulation and plant pathogen resistance.

To explore the molecular theory of stress resistance in pears, RT-qPCR has become a powerful and widely used method for analyzing and quantifying the related gene expression levels [17]. Before normalizing the RT-qPCR data, the stability of the reference genes under study must be tested. The ideal reference gene is one that is consistently and steadily expressed in all tissues under various conditions. Commonly used reference genes in plant research are the housekeeping genes, such as *ACT* (actin), *UBQ* (ubiquitin), *EF1α* (elongation factor 1 alpha), and *GAPDH* (glyceraldehyde-3-phosphate dehydrogenase) [18, 19]. However, they have a relatively stable expression only in some conditions in plants, and no single gene is consistently expressed under all kinds of stressful conditions [20]. In previous studies on pears, reference genes *PbGAPDHc1* and *PbGAPDHc2* were found to be more suitable than others for Yali (*Pyrus bretschneideri*) pears [21]. *TUB* (tubulin) and *WDP* (WD-repeat protein) were suitable reference genes in Dangsansu (*Pyrus bretschneideri*) pear leaves studied under cold conditions, while *UBQ* was expressed stably under the stresses of heat and salinity [22]. However, to the best of our knowledge, under biotic conditions, the expression of reference genes in pears is still uncertain and no invariable internal control gene has been reported in tissue stems cultured in an MS medium. Obviously, the lack of validated stably expressed reference genes affects the judgment of the functional study of genes.

In this study, we screened and selected the commonly used housekeeping genes, including *EF1α* [23], *ACT*, *SKD1* (suppressor of K^+ transport growth defect) [24], *YLS8* (thioredoxin-like protein) [25], *UBQ* [26], and *GAPDH* [27]. In our previous study, *TUB* and *WDP* were expressed stably in the transcriptome sequencing of Huangguan after being subjected to *V. nashicola*. The above-mentioned genes are often constitutively expressed and are required for basic cellular functions. We exposed Huangguan leaves to various experimental treatments; for instance, heat, cold, water deficiency, salinity, ABA, ETH, and pathogens. Considering that gene expression levels can correspondingly change under different conditions, each reference gene should be experimentally tested. This work represents the first effort for selecting optimal reference genes for pear tissue stems cultured under biotic and abiotic stresses. The accuracy of RT-qPCR results greatly improved after the selection of the validated candidate reference genes. Our aim was to find the most stable reference gene under all the above-mentioned conditions.

2. Materials and Methods

2.1. Plant Materials and Treatments

Huangguan pears were grafted on a three-year-old duli pear tree (2021 stock) cultivated in the orchard of the College of Horticulture, Shanxi Agricultural University (Jinzhong, China). The pear scab pathogen, *V. nashicola*, was cultivated at 22 °C in darkness in a growth chamber at the College of Horticulture at Shanxi Agricultural University. In 2022, the 3rd to 5th leaves from the one-year-old shoots were collected for further use. In the experiments, six different treatments were used: (1) foliar spray with the MeJA solution (0.1 mM); (2) foliar spray with the SA solution (0.1 mM); (3) foliar spray with the ABA solution (0.1 mM); (4) foliar spray with the ETH solution (0.1 mM); (5) smear inoculation of leaves using pear scab spore suspension (for the biotic stress treatment of leaves, the pathogen *V. nashicola* was washed with sterile distilled water and a spore suspension was prepared at a concentration of 1.0×10^5 CFU/mL); (6) and sterile distilled water (control). Three healthy shoots were used for replication. For each treatment, three biological replications were used. The sampled leaves were collected at 0, 12, 24, and 48 h post-treatment (hpt), instantly frozen in liquid nitrogen, and stored at −80 °C until further processing.

Abiotic stress assays were carried out with Huangguan leaves from buds regenerated in vitro. The tissue culture and the regeneration system of Huangguan were established as before [28]. One-year-old shoots were selected and cut into sections. Each section with a bud was sterilized for 90 s in 75% alcohol, followed by 8–10 min of periodic agitation in

0.1% $HgCl_2$. After sterilization, the materials were rinsed with sterilized water three times. Disinfected sections were placed on culture media containing the complete Murashige and Skoog (MS) medium with vitamins (Coolabor, Beijing, China), 6-BA (2 mg/L), and indole butyric acid (IBA) (0.5 mg/L). After buds sprouted, tissue culture stems were transferred to new MS media. The medium plus 40% (*w*/*v*) polyethylene glycol (PEG6000) simulated water-deficit stress, while the medium plus 400 mM NaCl simulated salinity stress. The sections were placed in treatments simulating 40 °C and 4 °C hot and cold stress, respectively. The controls grew at 23 ± 2 °C under cool white fluorescent lamps, which provided light in a 16 h photoperiod of 110 μmol/m^2 s. At 0, 48, and 96 hpt, the leaves of three replication shoots were collected and frozen in liquid nitrogen. A total of three biological replications were prepared for further research.

2.2. RNA Isolation and cDNA Synthesis

Total RNA was extracted from leaves using the CTAB-LiCl method [29], and the optical density (OD) was measured with a NanoDrop ND-1000 spectrophotometer using a UV spectrophotometer (Themo, USA) and gel electrophoresis. After running the RNA on an agarose gel, we used the RNA with an OD_{260}/OD_{280} ratio between 1.8 and 2.0, an OD_{260}/OD_{230} ratio greater than 2.0, and clear bands for 28S and 18S for cDNA synthesis. Subsequently, cDNA was synthesized following the instructions of the PrimeScript™ RT reagent Kit with gDNA Eraser (Perfect Real Time) kit (TaKaRa, Dalian, China). The resulting cDNA was stored at −20 °C.

2.3. Verification of Candidate Reference Genes

As previously reported, 8 candidate reference genes were selected and the primers were redesigned in primer-BLAST (https://www.ncbi.nlm.nih.gov/tools/primer-blast/index.cgi?LINK_LOC=BlastHome (accessed on 6 November 2021)) as per the following criteria: primer length 17–20 bp and GC content 40–60% (Table S1). All primers were synthesized by Sangon Biotech Co., Ltd. (Shanghai, China). Gel electrophoresis was used to confirm one PCR product per primer pair. Then, the PCR products were purified using the QIAquick Gel Extraction Kit (Qiagen, Hilden, North Rhine-Westphalia, Germany) and linked to pMD19-T (TaKaRa) transformed in DH5α *Coli*. Each clone of a gene was sequenced in Sangon Company. The accuracy of the sequences of candidate reference genes as query sequences was confirmed against pear genomes using BLASTN and BLASTX in NCBI (https://www.ncbi.nlm.nih.gov/ (accessed on 6 November 2021)).

2.4. Optimization of qPCR Conditions

As demonstrated by Zhao [3], qPCR conditions, such as different annealing temperatures (56, 58, 60, and 62 °C) and different primer concentrations (250, 300, 350, and 400 mM), were optimized for each factor. In each instance, the optimal factor was the one that had the lowest Cq value. Then, we used the optimal annealing temperature and the primer concentration for each primer pair and serial dilutions of the cDNA (1:10, 1:20, 1:40, 1:80, and 1:160 dilutions) to obtain the standard cDNA concentration curve on a logarithmic scale. For each primer, we calculated the standard curve and the PCR efficiency (E, %). We plotted the averaged Cq values from three technical replicates against lg (cDNA ng/reaction), as the initial cDNA concentration could be calculated by dividing 1000 ng of RNA by 10 μL of cDNA (i.e., 100 ng/μL) and the serially diluted cDNA (1:10, 1:20, 1:40, 1:80, and 1:160 dilutions) could be 10, 5, 2.5, 1.25, and 0.625 ng/μL. The correlation coefficients (R^2) and the standard curves equations of 8 candidate reference genes in pear leaves were calculated in Excel (Microsoft Office 2010). The amplification efficiencies (E) were calculated as $E\% = (10^{-1/A} - 1) \times 100$ using the slopes of the standard curves. Only primers with an amplification efficiency near 100% and a correlation coefficient (R^2) > 0.99 were used for further RT-qPCR amplification.

2.5. RT-qPCR and Data Analyses

For each qPCR reaction, we used a PCR reaction volume of 10 μL, which contained 5 μL of Green qPCR MasterMix (Biomed, China), 0.3 μL of each pair of primers, 1 μL of diluted cDNA, and 3.4 μL of ddH_2O. The cycling protocol was used as follows: 2 min at 95 °C followed by 40 rounds of 15 s at an annealing temperature of 95 °C and 40 rounds of 30 s at an annealing temperature of 60 °C. To verify the specificity of primer amplification, we generated a melting curve (95 °C for 10 s, 65 °C for 60 s and, 97 °C for 1 s). To monitor possible sampling error and experimental error, we used three biological replicates, with three technical replicates for each sample. When the programs were completed, we immediately calculated background-corrected fluorescence data and quantification cycle (Cq) values using LightCycler ®96 instrument software version 1.1.0.1320.

To assess and rank the feasibility of candidate reference genes, we analyzed the Cq of RT-qPCR in the web tool RefFinder (http://blooge.cn/RefFinder/ (accessed on 6 November 2021)) using GeNorm, NormFinder, and BestKeeper [30]. The outcome values indicated the stability of the reference genes. Using GeNorm (https://seqyuan.shinyapps.io/seqyuan_prosper (accessed on 6 November 2021)) [31], we ranked candidate genes according to their expression stability by calculating the M-values. The default value of $V_{n/n+1}$ was 0.15, which was slightly adjustable. If $V_{n/n+1}$ was no greater than 0.15, the optimal number of reference genes was n, indicating that the use of n most stable candidate genes as the reference genes is adequate for ensuring the accurate normalization of RT-qPCR data and the addition of one more candidate gene will not make a significant difference. Finally, we used RefFinder to obtain a comprehensive stability ranking of all the candidate reference genes.

2.6. Validation of Housekeeping Genes

A typical event in plants under abiotic and biotic stresses is a burst of reactive oxygen species (ROS), leading to a hypersensitive response and programmed cell death [32]. Plant NADPH oxidase, well known as a respiratory burst oxidase homolog (RBOH), mainly catalyzes the production of ROS [33]. Most plants have been found to respond to hormone and abiotic stresses [34,35]. Thus, to validate the reliability of the selected reference genes, we selected and investigated the expression levels of *PbRBOHC(Pbr038667.1)* as an object gene under various experimental conditions. For each stress, the relatively best and worst reference genes were selected for the experimental condition comparison study. The relative expression level of each housekeeping gene was calculated using the formula $L = 2^{-\Delta\Delta C_q}$.

2.7. Statics Analysis

We acquired raw fluorescence Cq values from qPCR experiments and analyzed them in Excel (Microsoft Office 2010). We created the figures in Origin 9.0 software. For each biological replicate of each tissue sample, we calculated the average Cq value of the three technical replicates. We calculated the coefficient of variation (CV, %) of the Cq values of each candidate gene for different samples individually and in combination. In addition, we calculated the correlation coefficients (R^2) and the standard curves equations of the 8 candidate reference genes in pear leaves in Excel (Microsoft Office 2010).

3. Results

3.1. Assessment of Primer Specificity and PCR Amplification Efficiency

We selected eight common reference genes *ACT, SKD1, YLS8, WDP, TUB, UBQ, GAPDH,* and *EF1α* and then redesigned specific primers using PCR amplicons of similar lengths (Table S1). Agarose gel electrophoresis showed a single band of the expected size (Figure S1). After sequencing, we compared the candidate reference genes in NCBI and confirmed the accuracy of these sequences. The cDNA sequences with redesigned primers of the candidate reference genes were consistent with their homologous sequences in pears. Afterward, we optimized the annealing temperature, the primer concentration, and the optimal template cDNA concentration range for each gene (Table S1). The melt curves of the eight candidate reference genes showed a single peak in each case, representing the specific primers (Figure S2). Under optimal steps, cDNA templates of the same leaf at dilutions of 1:10, 1:20, 1:40, 1:80, and 1:160 were subjected to qPCR and the standard concentration curve at a logarithmic scale was obtained for each primer pair. The standard curves equations of different reference genes were successfully calculated; the correlation coefficients (R^2) varied between 0.9949 and 0.9999. The amplification efficiencies (E) were calculated as E% = $(10^{-1/A} - 1) \times 100$ using the slopes (A) of the standard curves and maintained at 100 ± 10% (Table 1; Figure S3).

Table 1. List of genes and the optimized qPCR results for eight candidate reference genes.

Gene	Sequence ID	Amplicon Length (bp)	Annealing Tm (°C)	R^2	Amplification Efficiency (%)
ACT	GU830958	139	60	0.9999	104.39
SKD1	XM009366023	127	60	0.9992	106.66
YLS8	XM009353468	119	60	0.9993	111.02
WDP	Pbr038544	85	60	0.9995	111.29
TUB	Pbr042345	114	60	0.9949	111.62
UBQ	Pbr008614	88	60	0.9977	108.20
GAPDH	Pbr036263	118	60	0.9983	111.71
EF1α	XM_009376302	91	60	0.9993	108.39

3.2. *Transcript Abundance of the Eight Candidate Reference Genes in Pears under Hormone Treatments*

Time–course analysis of the average quantification cycle (Cq) values for each gene under different hormone treatments at 0–48 hpt ranged from 20.12 to 28.40 (Figure 1; Table S2). The coefficients of variation (CVs, %) of the Cq values of the eight genes (Table S2) at the same time points were ranked as *YLS8*(1.02) < *WDP*(2.85) < *SKD1*(2.89) < *TUB*(3.55) < *ACT*(3.88) < *UBQ*(6.49) < *EF1α*(6.65) < *GAPDH*(7.78), indicating that YLS8 had the most stable transcript abundance under mock treatment (Table S2). Under stress caused by SA/ETH/ABA hormones, *SKD1* (5.88/2.37/4.89) had the most stable transcript abundance, while *GAPDH* (1.80) had more stable expression than the others under MeJA hormone treatment.

3.3. *Analysis of the Expression Stability of Reference Genes under Hormone Treatments*

It was critical to analyze the expression stability of the candidate reference genes under different experimental conditions using statistical algorithms. We used BestKeeper, NormFinder, and GeNorm for the analysis.

3.3.1. BestKeeper Analysis

BestKeeper calculates the index value from the geometric mean of the Cq of each gene. It performs a Pearson correlation of each reference gene with the BestKeeper index to indicate the correlation of that gene with the index [36]. In BestKeeper, the smaller the values of the SD and the coefficient of variation, the higher the expression stability. In this study, if the SD of a particular gene was >1, it was excluded from the list of candidate reference genes and thus BestKeeper identified five stable reference genes among the eight under mock treatment. Only *SKD1* gene had a stable expression under SA, ETH, and ABA hormone treatment, and *GAPDH* was the most stable reference gene under MeJA treatment (Table 2).

3.3.2. NormFinder Analysis

For each candidate gene, NormFinder can provide a stability value that is a direct measurement of expression variation [37]. The most stably expressed gene had the smallest value. Hence, ACT had the most stable expression under mock and ABA treatments. Specifically, ACT ranked among the top three stable reference genes under other hormone treatments. Under SA, MeJA, and ETH hormone treatments, WDP, TUB, and UBQ were the most stably expressed genes, respectively (Table 2).

Table 2. Ranking of the expression stability of the eight candidate reference genes in pears under control conditions and hormone treatments with Cq values being calculated using statistical algorithm software.

Mock		SA		MeJA		ETH		ABA	
Gene	Value	Gene	Value	Gene	Value	Gene	Value	Gene	Value
Calculated in BestKeeper									
YLS8	0.17	*SKD1*	0.91	*GAPDH*	0.36	*SKD1*	0.37	*SKD1*	0.76
WDP	0.45	*WDP*	1.08	*EF1α*	0.62	*WDP*	0.99	*TUB*	1.07
SKD1	0.51	*YLS8*	1.16	*ACT*	0.68	*TUB*	1.17	*WDP*	1.09
TUB	0.63	*ACT*	1.18	*TUB*	0.94	*YLS8*	1.35	*YLS8*	1.36
ACT	0.63	*TUB*	1.37	*UBQ*	0.95	*UBQ*	1.75	*UBQ*	1.63
EF1α	1.11	*EF1α*	1.54	*WDP*	1.08	*ACT*	2.00	*ACT*	2.00
UBQ	1.33	*UBQ*	1.68	*SKD1*	1.47	*GAPDH*	2.39	*GAPDH*	2.44
GAPDH	1.40	*GAPDH*	1.82	*YLS8*	1.50	*EF1α*	2.60	*EF1α*	2.77
Calculated in NormFinder									
ACT	0.22	*WDP*	0.09	*TUB*	0.33	*UBQ*	0.28	*ACT*	0.48
WDP	0.27	*ACT*	0.16	*ACT*	0.35	*YLS8*	0.30	*YLS8*	0.52
EF1α	0.56	*EF1α*	0.43	*UBQ*	0.44	*ACT*	0.43	*UBQ*	0.53
TUB	0.70	*TUB*	0.51	*WDP*	1.09	*WDP*	0.71	*WDP*	0.82
YLS8	0.74	*UBQ*	0.68	*YLS8*	1.09	*TUB*	0.77	*TUB*	0.87
SKD1	0.77	*SKD1*	0.84	*GAPDH*	1.27	*GAPDH*	1.05	*GAPDH*	1.18
UBQ	0.94	*YLS8*	0.88	*SKD1*	1.37	*SKD1*	1.72	*SKD1*	1.65
GAPDH	0.96	*GAPDH*	0.90	*EF1α*	1.44	*EF1α*	1.94	*EF1α*	1.89
Calculated in GeNorm									
SKD1 | *TUB*	0.19	*ACT* | *WDP*	0.31	*ACT* | *UBQ*	0.69	*WDP* | *TUB*	0.40	*YLS8* | *WDP*	0.38
WDP	0.31	*TUB*	0.47	*TUB*	0.72	*YLS8*	0.61	*TUB*	0.62
YLS8	0.43	*EF1α*	0.56	*YLS8*	0.86	*UBQ*	0.69	*UBQ*	0.87
ACT	0.52	*UBQ*	0.63	*SKD1*	0.98	*ACT*	0.81	*ACT*	0.98
EF1α	0.68	*YLS8*	0.72	*WDP*	1.08	*GAPDH*	0.94	*SKD1*	1.09
UBQ	0.82	*SKD1*	0.78	*GAPDH*	1.25	*SKD1*	1.13	*GAPDH*	1.23
GAPDH	0.88	*GAPDH*	0.84	*EF1α*	1.34	*EF1α*	1.35	*EF1α*	1.42

3.3.3. GeNorm Analysis

As per GeNorm analysis, the cut-off range of stability value is less than 0.5 [38]. So, in terms of gene expression, the gene with the lowest value is considered to be the most stable reference gene and vice versa. As per GeNorm analysis, *SKD1* and *TUB* were the most stably expressed genes under mock treatment, *ACT* and *WDP* were the most stably expressed genes under SA treatment, *ACT* and *UBQ* were the most stably expressed genes under MeJA treatment, *WDP* and *TUB* were the most stably expressed genes under ETH treatment, and *YLS8* and *WDP* were the most stably expressed genes under ABA treatment (Table 2). Meanwhile, the number of reference genes used for qPCR analysis relied on the value of $V_{n/n+1}$ ($n \geq 2$) (Figure 2). When the value of $V_{n/n+1}$ was less than 0.15, the optimal number of reference genes was n. Therefore, under mock treatment, the value of $V_{2/3}$ was lower than the default threshold of 0.15. So, the use of the two most stably expressed reference genes *SKD1* and *TUB* ensured accurate normalization of qPCR data. Adding one more reference gene did not make a significant difference. Since the value of $V_{3/4}$ (0.146) under SA treatment was lower than 0.15, it was practical to use three reference genes (*ACT*, *WDS*, and *TUB*) in qPCR. Although the values of all $V_{n/n+1}$ were higher than the default value under MeJA, ETH, and ABA treatments, this did not indicate the number of reference genes used for normalization in RT-qPCR analysis.

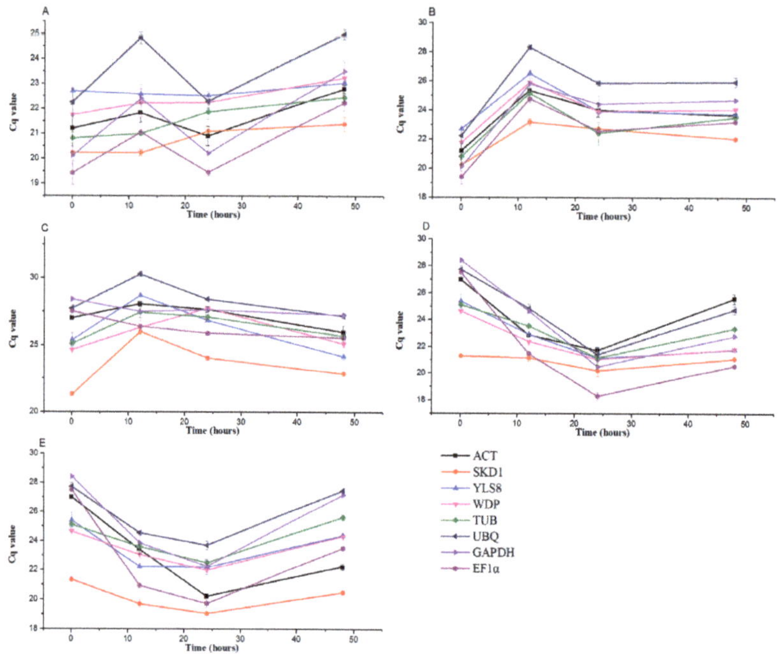

Figure 1. Transcript abundance of the eight reference genes in pear leaves under control conditions and when exposed to different hormones as measured using qPCR. (**A**) Mock, (**B**) SA, (**C**) MeJA, (**D**) ETH, and (**E**) ABA treatment.

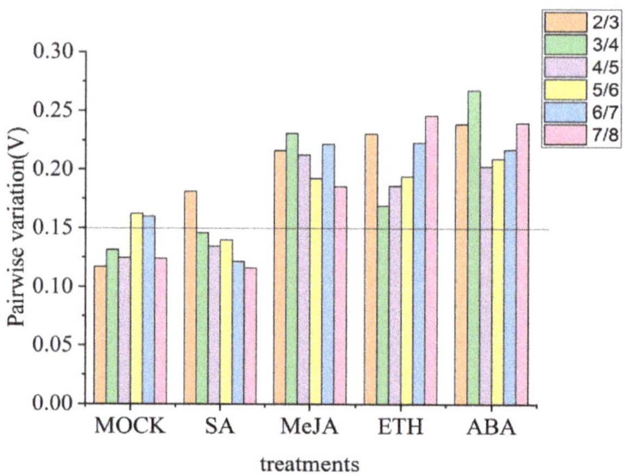

Figure 2. Analysis of reference genes for qPCR normalization using GeNorm in pear leaves under control condition and different hormones. The pairwise variation ($V_{n/n+1}$) in the eight candidate reference genes under different treatments was calculated using GeNorm software, and the optimal number of reference genes was determined using the lowest number of genes with a $V_{n/n+1}$ value smaller than 0.15.

3.3.4. RefFinder Analysis

Slight differences in the analysis results of the three software programs are clearly visible. The results were confirmed using the web-based tool RefFinder, which compared and re-ranked the tested reference genes on the basis of the geometric mean of the weights of the three software programs. As per the comprehensive ranking performed by RefFinder, the stability rankings of the genes under mock treatment were *WDP* > *ACT* > *TUB* > *SKD1* > *YLS8* > *EF1α* > *UBQ* > *GAPDH*, and the stability rankings of the genes under each hormone treatment were as follows: *WDP* > *ACT* > *EF1α* > *TUB* > *SKD1* > *UBQ* > *YLS8* > *GAPDH* under SA treatment; *ACT* > *TUB* > *UBQ* > *GAPDH* > *YLS8* > *WDP* > *EF1α* > *SKD1* under MeJA treatment; *UBQ* > *WDP* > *YLS8* > *TUB* > *ACT* > *SKD1* > *GAPDH* > *EF1α* under ETH treatment; and *YLS8* > *WDP* > *ACT* > *TUB* > *UBQ* > *SKD1* > *GAPDH* > *EF1α* under ABA treatment (Table 3). To sum up, *WDP* and *ACT* exhibited the best stability under hormone treatments for normalization in qPCR.

Table 3. Rankings of the expression stability of the eight candidate reference genes in pears under control conditions and hormone treatments, with Cq values being calculated in RefFinder.

Mock		SA		MeJA		ETH		ABA	
Gene	Value	Gene	Value	Gene	Value	Gene	Value	Gene	Value
WDP	1.86	*WDP*	1.19	*ACT*	1.57	*UBQ*	2.11	*YLS8*	1.68
ACT	2.66	*ACT*	2.00	*TUB*	2.21	*WDP*	2.38	*WDP*	2.63
TUB	2.83	*EF1α*	3.83	*UBQ*	2.59	*YLS8*	2.63	*ACT*	2.78
SKD1	3.08	*TUB*	3.94	*GAPDH*	3.98	*TUB*	2.94	*TUB*	3.50
YLS8	3.31	*SKD1*	3.98	*YLS8*	5.03	*ACT*	4.05	*UBQ*	3.66
EF1α	4.24	*UBQ*	5.44	*WDP*	5.18	*SKD1*	4.30	*SKD1*	4.14
UBQ	7.00	*YLS8*	5.45	*EF1α*	5.66	*GAPDH*	6.24	*GAPDH*	6.48
GAPDH	8.00	*GAPDH*	8.00	*SKD1*	6.44	*EF1α*	8.00	*EF1α*	8.00

3.4. Transcript Abundance of the Eight Candidate Reference Genes in Pears under Abiotic and Biotic Stress Treatments

To select the best reference gene in pears under abiotic and biotic stresses, the leaves of buds were cultivated in MS media under 4 °C, 40 °C, 40% PEG-6000, and 400 mM NaCl. We examined the expression of eight reference genes under different abiotic and biotic stresses. The transcript abundance of each gene in terms of the same amount of total RNA per reaction was measured in the leaves, and the average Cq values were from 22.42 to 29.35 (Figure 3; Table S2). Lower Cq values reflected higher mRNA transcript levels, and *SKD1* had the higher transcript levels among the reference genes under cold (23.42), heat (22.42), and salinity (23.76) stresses. *YLS8* (25.16), *UBQ* (28.39), and *ACT* (27.6) had the lowest expression levels under cold, heat, and salinity stresses conditions, respectively. In pear leaves under water-deficit stress, *TUB* (25.16) had a high expression level, while *ACT* (27.60) had the lowest expression level. According to the coefficients of variation (CVs, %) of the Cq values, at the three time points, the eight genes were ranked as follows under various stresses: *UBQ* < *GAPDH* < *EF1α* < *SKD1* < *TUB* < *WDP* < *YLS8* < *ACT* under cold stress, *ACT* < *GAPDH* < *UBQ* < *SKD1* < *YLS8* < *WDP* < *EF1α* < *TUB* under heat stress, *UBQ* < *GAPDH* < *TUB* < *SKD1* < *EF1α* < *WDP* < *YLS8* < *ACT* under water-deficit stress, and *GAPDH* < *SKD1* < *ACT* < *WDP* < *TUB* < *YLS8* < *EF1α* < *UBQ* under salinity stress (Table S2). Accordingly, *UBQ*, *ACT*, and *GAPDH* were stably expressed genes used in this aspect of qPCR analysis under abiotic stress.

Next, we studied the expression stability of each candidate reference genes in Huangguan pear leaves under biotic treatment with *V. nashicola*, the causal agent of scab disease. Time-course analysis of the average Cq values for each gene at 0, 12, 24, and 48 hpt showed that *EF1α* (19.41) had the highest expression level and *UBQ* (27.58) had the lowest expression level among the eight candidate reference genes in leaves after infection (Figure 3). The coefficients of variation (CVs, %) of the Cq values of the eight genes across the four time points were ranked as *SKD1* < *YLS8* < *WDP* < *ACT* < *TUB* < *UBQ* < *EF1α* < *GAPDH* (Table S2).

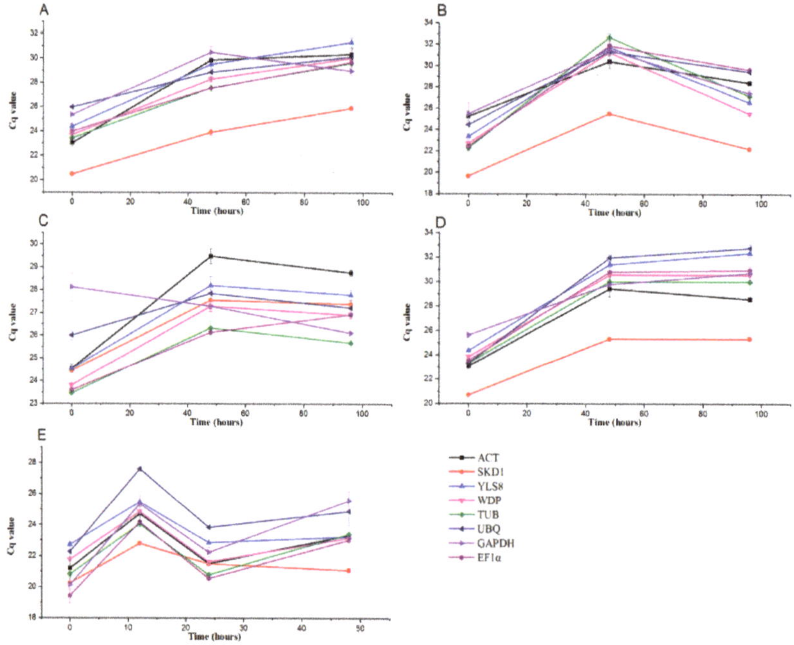

Figure 3. Transcript abundance of the eight reference genes of pear leaves under abiotic and biotic stresses measured using RT-qPCR. Cq values indicate the mean values of transcript abundance from three independent replicates ± standard errors (vertical bars). (**A**) Cold, (**B**) heat, (**C**) water deficit, (**D**) salinity, and (**E**) *V. nashicola*.

3.5. Analysis of the Expression Stability of Reference Genes under Abiotic and Biotic Stress Treatment

It is critical to analyze the expression stability of the candidate reference genes under abiotic and biotic stress conditions using statistical algorithms. Under the optimized qPCR conditions, the transcript abundance of the eight candidate reference genes were quantified in leaves subjected to cold, heat, PEG, NaCl, and *V. nashicola*.

3.5.1. BestKeeper Analysis

According to the same standard, the transcript abundance of the eight candidate reference genes in Huangguan pear leaves under abiotic and biotic stresses were analyzed using the statistical algorithm software BestKeeper. Because the value of stably expressed genes had to be <1, many genes in analysis results would be omitted under abiotic stresses. *UBQ* (0.68) and *GAPDH* (0.71) were stable as reference genes under water-deficit stress (Table 3). *YLS8* (0.87) and *WDP* (0.91) were stable as reference genes in leaves infected with *V. nashicola* (Table 4).

3.5.2. NormFinder Analysis

We used NormFinder to calculate the value of the eight reference genes according to the Cq of qPCR. In the stability ranking of Huangguan leaves under four abiotic stresses, *WDP* (0.71) and *YLS8* (0.86) were more stable than the other genes. For each abiotic stress, *GAPDH* was the least stable reference gene. *WDP* (0.11) was the best one in cold-stressed leaves. Under heat treatment, *YLS8* (0.67) was the most stable reference gene. In leaf samples subjected to PEG or NaCl, *SKD1* (0.09) and *TUB/WDP* (0.01) were the best choices, respectively (Table 4). In leaf samples subjected to the pathogen *V. nashicola*, *ACT* (0.28) and *WDP* (0.31) were the top two stable reference genes.

Table 4. Ranking of the expression stability of the eight candidate reference genes in pears under different abiotic and biotic stresses, with Cq values being calculated using statistical algorithm software.

Cold		Heat		Water Deficit		Salinity		*V. nashicola*	
Gene	Value	Gene	Value	Gene	Value	Gene	Value	Gene	Value
Calculated using BestKeeper									
UBQ	1.56	*UBQ*	1.56	*UBQ*	0.68	*SKD1*	2.06	*SKD1*	0.74
GAPDH	1.95	*GAPDH*	1.95	*GAPDH*	0.71	*GAPDH*	2.06	*YLS8*	0.95
SKD1	1.97	*SKD1*	1.97	*TUB*	1.13	*ACT*	2.63	*WDP*	1.16
EF1α	2.06	*EF1α*	2.06	*EF1α*	1.30	*TUB*	2.97	*ACT*	1.33
TUB	2.30	*TUB*	2.30	*SKD1*	1.34	*WDP*	2.99	*TUB*	1.46
WDP	2.43	*WDP*	2.43	*WDP*	1.46	*EF1α*	3.26	*UBQ*	1.59
YLS8	2.68	*YLS8*	2.68	*YLS8*	1.54	*YLS8*	3.33	*EF1α*	1.80
ACT	3.14	*ACT*	3.14	*ACT*	2.03	*UBQ*	4.01	*GAPDH*	2.14
Calculated using NormFinder									
WDP	0.11	*YLS8*	0.67	*SKD1*	0.09	*TUB*	0.01	*ACT*	0.15
TUB	0.32	*SKD1*	0.92	*TUB*	0.18	*WDP*	0.01	*WDP*	0.42
EF1α	0.47	*UBQ*	0.96	*WDP*	0.29	*EF1α*	0.38	*TUB*	0.44
SKD1	0.49	*WDP*	1.04	*YLS8*	0.47	*YLS8*	0.64	*EF1α*	0.55
YLS8	0.60	*GAPDH*	1.05	*UBQ*	0.48	*ACT*	0.66	*UBQ*	0.72
UBQ	0.98	*ACT*	1.36	*EF1α*	0.62	*SKD1*	1.24	*YLS8*	0.85
ACT	1.34	*TUB*	1.56	*ACT*	1.33	*GAPDH*	1.28	*SKD1*	1.06
GAPDH	1.63	*EF1α*	1.88	*GAPDH*	2.61	*UBQ*	1.63	*GAPDH*	1.41
Calculated using GeNorm									
SKD1 | *EF1α*	0.14	*YLS8* | *WDP*	0.29	*YLS8* | *WDP*	0.10	*WDP* | *TUB*	0.03	*ACT* | *WDP*	0.30
TUB	0.29	*GAPDH*	0.91	*SKD1*	0.19	*EF1α*	0.26	*TUB*	0.40
WDP	0.37	*SKD1*	1.00	*TUB*	0.32	*YLS8*	0.41	*EF1α*	0.57
YLS8	0.48	*ACT*	1.17	*EF1α*	0.44	*ACT*	0.60	*UBQ*	0.70
UBQ	0.66	*UBQ*	1.27	*UBQ*	0.61	*SKD1*	0.83	*YLS8*	0.78
ACT	0.89	*TUB*	1.45	*ACT*	0.74	*GAPDH*	0.95	*SKD1*	0.87
GAPDH	1.11	*EF1α*	1.63	*GAPDH*	1.22	*UBQ*	1.13	*GAPDH*	1.02

3.5.3. GeNorm Analysis

The expression stability of the eight candidate genes was further analyzed using GeNorm software. According to GeNorm analysis, the cut-off range of the stability value is < 1.5 (Table 4) but when choosing a sample of reference genes on the basis of $V_{n/n+1}$ (Figure 4), the V of $V_{2/3}$ was less than the threshold of 0.15 for all eight genes in cold, water-deficit, salinity, and pathogen-stressed leaves. So, we referred to *SKD1* | *EF1α* (0.14), *YLS8* | *WDP* (0.10), *WDP* | *TUB* (0.03), and *ACT* | *WDP* (0.30) as housekeeping genes. Although the transcript abundance of housekeeping genes maintained the GeNorm analysis condition, none of the values of $V_{n/n+1}$ under heat stress were less than 0.15, which means that we were uncertain about the number of reference genes to use in qPCR. From an economic perspective, the top two genes *YLS8* | *WDP* were acceptable reference genes.

3.5.4. RefFinder Analysis

RefFinder can provide a stability value that is a direct measurement of expression variation by comparing and re-ranking the results of three computer programs. In a comprehensive ranking, *WDP* was among the top three most stable reference genes under various abiotic stresses. *WDP*, *YLS8*, *SKD1*, and *TUB* were the best reference genes under cold, heat, water-deficient, and salinity stresses, respectively (Table 5). Under biotic stress inflicted by *V. nashicola*, the stability rankings of the genes were *ACT* > *WDP* > *TUB* > *SKD1* > *YLS8* > *EF1α* > *UBQ* > *GAPDH*.

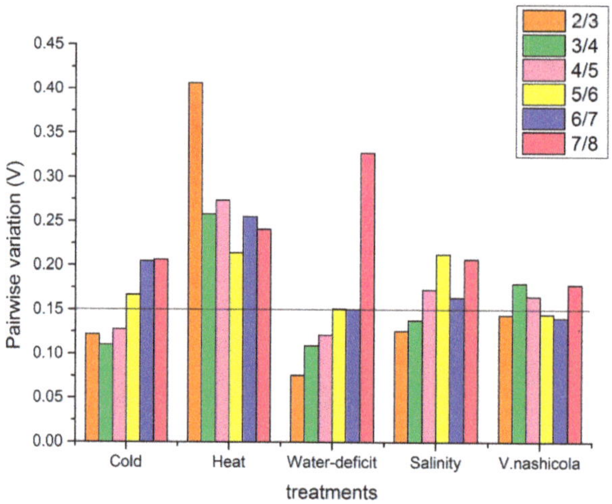

Figure 4. Analysis of reference genes for qPCR normalization using GeNorm in pear leaves under different stresses. The pairwise variation ($V_{n/n+1}$) in the eight candidate reference genes under abiotic and biotic treatments was calculated using GeNorm software, and the optimal number of reference genes was determined using the lowest number of genes with a $V_{n/n+1}$ value smaller than 0.15.

Table 5. Ranking of the expression stability of the eight candidate reference genes in pears under different abiotic and biotic stresses, with Cq values being calculated in RefFinder.

Cold		Heat		Water Deficit		Salinity		*V. nashicola*	
Gene	Value	Gene	Value	Gene	Value	Gene	Value	Gene	Value
WDP	2.21	*YLS8*	1.50	*SKD1*	1.97	*TUB*	1.68	*ACT*	1.41
EF1α	2.45	*SKD1*	2.38	*WDP*	2.45	*WDP*	1.78	*WDP*	1.86
SKD1	2.63	*WDP*	3.13	*TUB*	2.91	*EF1α*	3.57	*TUB*	3.41
TUB	2.78	*GAPDH*	3.41	*YLS8*	3.25	*SKD1*	3.83	*SKD1*	4.30
UBQ	3.83	*ACT*	3.66	*UBQ*	3.66	*ACT*	4.40	*YLS8*	4.36
YLS8	5.44	*UBQ*	4.36	*EF1α*	4.95	*YLS8*	4.60	*EF1α*	4.60
GAPDH	5.66	*TUB*	7.00	*GAPDH*	5.66	*GAPDH*	5.12	*UBQ*	5.48
ACT	7.24	*EF1α*	8.00	*ACT*	7.24	*UBQ*	8.00	*GAPDH*	8.00

3.5.5. Validation of Selected Reference Genes

To normalize the expression level of *PbRBOHC*, we used the most and least stably expressed reference genes under various conditions on the basis of the analysis of the statistical algorithms above. In other words, the relative expression level (Q) of *PbRBOHC* was calculated using the formula $Q = 2^{-\Delta\Delta C_t}$. The genes were analyzed after mock (*WDP* and *GAPDH*), SA (*WDP* and *GAPDH*), MeJA (*ACT* and *SKD1*), ETH (*UBQ* and *EF1α*), and ABA (*YLS8* and *EF1α*) treatment and under *V. nashicola* (*ACT, GAPDH*), cold (*WDP, ACT*), heat (*YLS8, EF1α*), water-deficit (*SKD1, ACT*), and salinity (*TUB, UBQ*) conditions. Two reference genes were used to normalize the expression level of *PbRBOHC* and *PbRBOHC* was found to be expressed differently in comparison with the two reference genes (Figure 5). For instance, compared with the stable reference gene *GAPDH* and the unstable reference gene *WDP* in the mock treatment (Figure 5A), the expression of the target gene *PbRBOHC* decreased and was lower. In SA treatment (Figure 5B), it was more highly expressed compared with the unstable reference gene. In some other conditions, the expression pattern of *PbRBOHC* was totally different. After ABA (Figure 5E) treatment, *PbRBOHC* remained almost unchanged when normalized with the stable reference gene *YLS8*. However, its expression level rose with the unstable *EF1α*. The target gene was highly expressed and had a time–effect expression pattern under *V. nashicola* (Figure 5F), cold (Figure 5G), heat

(Figure 5H), water-deficit (Figure 5I), and salinity (Figure 5J) treatments when normalized with stable reference genes but hardly showed these trends when normalized with unstable ones. Further examination of the role of *PbRBOHC* in pear leaves revealed that its expression levels increased under *V. nashicola*, cold, heat, water-deficit, and salinity treatments, which is possibly related to biotic and abiotic stress causing an ROS burst.

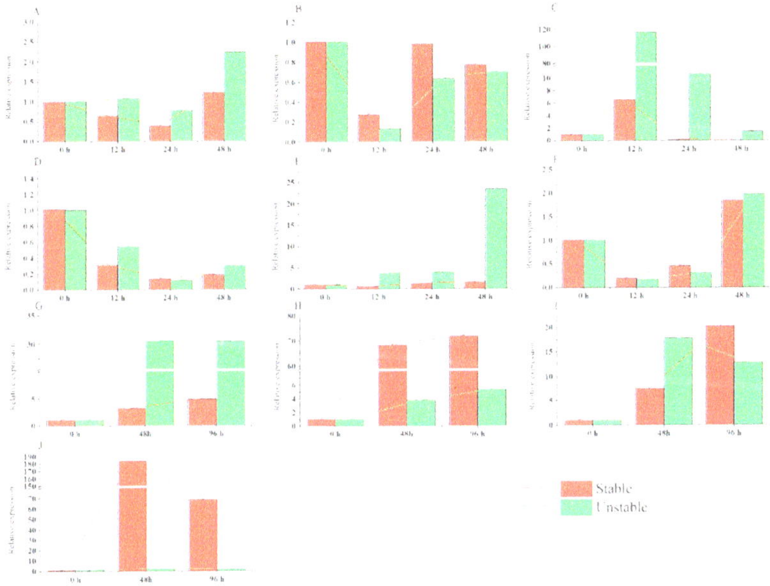

Figure 5. The calculated relative expression levels of *PbRBOHC* using the most and least stable reference genes in pear leaves under different experimental stresses. Error bars represent the SD. (**A**) Mock, (**B**) SA, (**C**) MeJA, (**D**) ETH, (**E**) ABA, and (**F**) *V. nashicola* treatment and (**G**) cold, (**H**) heat, (**I**) water-deficit, and (**J**) saline conditions.

4. Discussion

As a gene expression research tool, RT-qPCR is a convenient method of identifying gene expression and exploring the status and characteristics of plants. However, factors such as RNA quality and integrity, primer specificity, and concentration can influence the amplification efficiency and expression results of genes. To ensure equal qPCR amplification efficiency, it is necessary to optimize these factors for the relative quantification of the expression of reference and target genes [3]. In our study, we redesigned the primers of selected housekeeping genes to ensure the specificity of PCR products of less than 150 bp. A series of adjustments were made so that each candidate reference gene achieved $R^2 > 0.99$ and $E = 100 \pm 10\%$. Normalization of the RT-qPCR data required internal control genes that display uniform expression under different conditions. Otherwise, without a suitable referent gene, the resulting Cq values of qPCR could lead to inaccurate expressions of target genes [38]. NormFinder, GeNorm, and BestKeeper, three common pieces of software, are often used to evaluate the expressed Cq values to select the best reference gene under various conditions. Although in some instances, the ranking using BestKeeper software was almost the converse of rankings performed using other software, as is the case compared to previous studies [23,39,40], RefFinder can help to comprehensively analyze the values and overall rank of the reference genes. We applied our optimized protocol to test the expression stability of a set of reference genes in the leaves of Huangguan pears. The leaves were of particular interest for studying gene expression under the different hormone treatments or environmental stresses. Different statistical algorithms of the different analysis approaches provided different validation ranks in the same treatment. Therefore, we recommend using four pieces of software to ensure that the results are convincing.

According to previous studies on the selection of plant reference genes for RT-qPCR, the expression level of a reference gene might not be constant under various stress conditions in citrus fruits [41]. For example, Xu [42] suggested that in pears subjected to abiotic stresses, *ACT2/7*, ubiquitin extension protein (*UBI*), and yellow-leaf-specific gene 8 (*YLS8*) were the most stable housekeeping genes among 18 selected candidate reference genes. *ACT* and tubulin have been widely used as reference genes in citrus fruits [43] and peach [44] under abiotic stresses. In potato, *EF1α* and *sec3* were the most stably expressed genes under drought and osmotic stress conditions but *EF1α* and *APRT* were the most stable genes under cold stress [23]. However, in potato tubers, the most suitable reference genes were *C2*, *sec3*, and *CUL3A* [45]. In blueberry leaves and root, *GAPDH* was the worst reference gene under high alkalinity but *EF1α* was the best under PEG-simulated drought [46]. In our study, eight commonly used housekeeping genes were studied. We applied hormones to these genes because phytohormones are mainly related to biotic and abiotic stresses in plants. We also included various experimental stresses, which had a significant influence on the growth and quality of horticultural crops. Interestingly, we found some housekeeping genes were consistent in the three analysis programs. The results showed that *ACT* and *WDP* were relatively stable reference genes under hormones and biotic treatments. In abiotic stresses, *WDP* had more stable expression than *ACT* because *ACT* was ranked at the bottom for cold and water-deficit stresses. Although the ranking results were different, these differences revealed the expression levels of the invariable genes in pears under various conditions. In RefFinder analysis, the expression levels of *EF1α* and *GAPDH* were unstable under biotic and abiotic stresses. Chen [22] was the first to select WD-repeat protein (*WDP*) gene as a housekeeping gene from the transcriptome data [22]. It has been proved to be the most stable reference gene under cold treatment in different tissues of Dangshansu pear. After comparing 10 reference genes in pear pollen samples under ABA and abiotic stress treatments, especially high temperature and NaCl, Chen found that *PbrEF1α* was the most stable reference gene. After redesigning the primer and optimizing RT-qPCR, we also proved that WDP was among the top two best housekeeping genes under abiotic and biotic stresses in Huangguan, while *EF1α* ranked second in leaves only under cold-stress treatment in our study. This difference was in diverse organs or tissues because reference genes were unstable in different tissues.

The basic rule for a reference gene is that its expression level is insensitive to the exterior environment or any internal development changes. The Cq values of ideal reference genes among different samples were expected to be the same. However, according to our study, the Cq values varied from 19 to 30 (Figures 1 and 3) and the CV of the Cq value ranged from 1.02 to 17.88 (Table S2). It was hard to tell which reference gene had constant expression under all experimental conditions in pear samples (Figure 6) because some specific housekeeping genes were regulated under various exterior treatments. For instance, salinity and ABA treatments greatly affected *UBQ* and *EF1α* depolymerization and expression, which also occurred in a study on Arabidopsis and tobacco [47]. Through the study of the *PbRBOHC* expression level, different reference genes were used to normalize gene expression. The expression level of the target gene was inconsistent with the above reference genes. We also found that the analysis results of the gene expression pattern changed with different reference genes. Using an unvalidated reference gene to normalize led to errors. Thus, it was necessary and significant to validate a reliable reference gene to normalize gene expression profiling under certain conditions.

In summary, in this study, we analyzed the stability of reference genes for RT-qPCR in pear leaves subjected to different hormone treatments and various biotic and abiotic stresses. *WDP* was suitable for gene quantification in pear under various experimental stresses, except MeJA treatment. *ACT* had the most stable expression under MeJA. Moreover, the best reference gene was not always the most stable one under specific conditions. More candidate reference genes should be explored with the release of the pear genome sequences and abundant transcriptome sequencing [48], such as annexin and β-tubulin. The results obtained will provide guidelines for selecting reference genes and analyzing gene expression in pears under different experimental conditions. They will also provide a reference for the analysis of other crops via suitable RT-qPCR normalization conditions in the future.

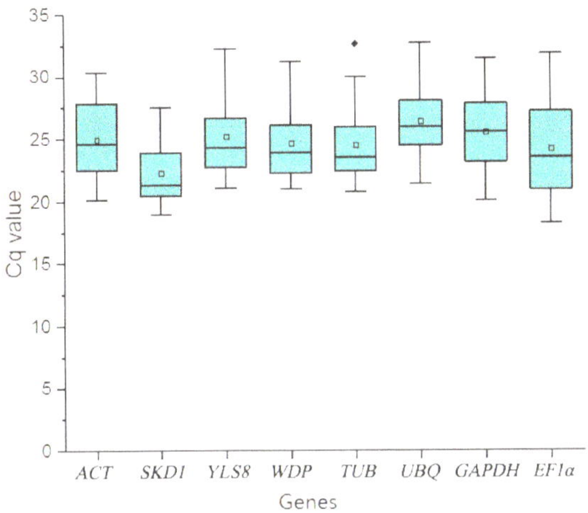

Figure 6. RT-qPCR Cq values of the candidate reference genes in pear leaves. Candidate reference genes were analyzed in all conditions. Boxes indicate the 25th and 75th percentiles. Lines across boxes represent the median values. Circles represent the mean values, and the spot represents the outlier. Whiskers represent the maximum and minimum values.

5. Conclusions

In this study, we valuated eight common candidate reference genes in pear (*Pyrus bretschneideri* Rehd.) leaves subjected to hormonal (SA, MeJA, ETH, and ABA), biotic (*V. nashicola*), and abiotic (cold, heat, water deficit, and salinity) stresses. As per the results, *WDP* was a relatively stable reference gene under all conditions except MeJA, while *ACT* had the most stable expression under MeJA. It is recommended that only *EF1α* and *GAPDH* be used as reference genes under experimental treatments.

Supplementary Materials: The following supporting information can be downloaded at https://www.mdpi.com/article/10.3390/horticulturae9020275/s1, Figure S1: The plot of the averaged Cq values of three technical replicates against the Log (cDNA in ng/reaction) for optimizing qPCR conditions for each candidate reference genes in pears; Figure S2: Figure S2. The analysis of specific primers and PCR efficiency for eight reference genes in pear leaves under control condition. Melting curves of reference genes showing single peaks. (a) *ACT*, (b) *SKD1*, (c) *YLS8*, (d) *WDP*, (e) *TUB*, (f) *UBQ*, (g) *GAPDH*, (h) *EF1α*; Table S1: Primer sequences used for qRT-PCR and the details of optimized protocol; Table S2: The Ct value of the reference genes of pear leaves under various experimental treatments.

Author Contributions: Conceptualization, P.Z., L.H., Y.W., X.L., X.F. and L.L.; investigation, Y.W. and X.L.; methodology, L.H.; validation and writing—original draft preparation, P.Z.; writing—review and editing, X.F.; project administration, L.L.; funding acquisition, L.L. All authors have read and agreed to the published version of the manuscript.

Funding: This research was financially supported by the Basic Research Program of Shanxi Province (Free Exploration) (grant no. YDZJSX2022A042).

Data Availability Statement: Not applicable.

Conflicts of Interest: The authors declare that they have no known competing financial interests in this paper.

References

1. Wu, J.; Wang, Z.W.; Shi, Z.B.; Zhang, S.; Ming, R.; Zhu, S.L.; Khan, M.A.; Tao, S.T.; Korban, S.S.; Wang, H.; et al. The genome of the pear (*Pyrus bretschneideri* Rehd.). *Genome Res.* **2013**, *23*, 396–408. [CrossRef] [PubMed]
2. Wang, Y.T.; Li, Y.; Sun, Y.H. The breeding and application of the middle season ripening, scab resistant new pear variety-Huangguan pear. *J. Hebei Agric. Sci.* **1998**, *2*, 40–42. [CrossRef]
3. Zhao, F.Z.; Maren, N.A.; Kosentka, P.Z.; Kosentka, P.Z.; Liao, Y.Y.; Lu, H.Y.; Duduit, J.R.; Huang, D.B.; Ashrafi, H.; Zhao, T.J.; et al. An optimized protocol for stepwise optimization of real-time RT-PCR analysis. *Hortic. Res.* **2021**, *8*, 179. [CrossRef] [PubMed]
4. Liu, Y.; Yang, T.Y.; Lin, Z.K.; Gu, B.J.; Xing, C.H.; Zhao, L.Y.; Dong, H.Z.; Gao, J.Z.; Xie, Z.H.; Zhang, S.L.; et al. A WRKY transcription factor *PbrWRKY53* from *Pyrus betulaefolia* is involved in drought tolerance and AsA accumulation. *Plant Biotechnol. J.* **2019**, *17*, 1770–1787. [CrossRef]
5. Xing, C.H.; Liu, Y.; Zhao, L.Y.; Zhang, S.L.; Huang, X.S. A novel MYB transcription factor regulates ascorbic acid synthesis and affects cold tolerance. *Plant Cell Environ.* **2019**, *42*, 832–845. [CrossRef]
6. Dong, H.Z.; Wang, C.M.; Xing, C.H.; Yang, T.Y.; Yan, J.X.; Gao, J.Z.; Li, D.L.; Wang, R.; Blumwald, E.; Zhang, S.L.; et al. Overexpression of *PbrNHX2* gene, a Na^+/H^+ antiporter gene isolated from *Pyrus betulaefolia*, confers enhanced tolerance to salt stress via modulating ROS levels. *Plant Sci.* **2019**, *285*, 14–25. [CrossRef]
7. Mizuno, A.; Tsukamoto, T.; Shimizu, Y.; Ooya, H.; Matsuura, T.; Saito, N.; Sato, S.; Kikuchi, S.; Uzuki, T.; Azegami, K. Occurrence of bacterial black shoot disease of European pear in Yamagata prefecture. *J. Gen. Plant Pathol.* **2010**, *76*, 43–51. [CrossRef]
8. Rou-Zi, A.; Wang, Q. Bionomics of grapholitha molesta busck and its control measures. *Biol. Disaster Sci.* **2018**, *41*, 160–162. [CrossRef]
9. Ishii, H.; Yanase, H. *Venturia Nashicola*, the scab fungus of Japanese and Chinese pears: A species distinct from *V. pirina*. *Mycol. Res.* **2000**, *104*, 755–759. [CrossRef]
10. Ma, M.Z. Analysis of harmful symptoms and epidemic causes of pear scab. *China Agric. Inf.* **2014**, *7*, 100. [CrossRef]
11. Gray, W.M. Hormonal regulation of plant growth and development. *PLoS Biol.* **2004**, *2*, E311. [CrossRef] [PubMed]
12. Kitagawa, Y.; Yamamoto, H.; Oritani, T. Biosynthesis of abscisic acid in the fungus *Cercospora cruenta* stimulation of biosynthesis by water stress and isolation of a transgenic mutant with reduced biosynthetic capacity. *Plant Cell Physiol.* **1995**, *36*, 557–564. [CrossRef]
13. Pré, M.; Atallah, M.; Champion, A.; De Vos, M.; Pieterse, C.M.J.; Memelink, J. The AP2/ERF domain transcription factor ORA59 integrates jasmonic acid and ethylene signals in plant defense. *Plant Physiol.* **2008**, *147*, 1347–1357. [CrossRef] [PubMed]
14. Weingart, H.; Ullrich, H.; Geider, K.; Völksch, B. The role of ethylene production in virulence of *Pseudomonas syringae* pvs. *glycinea* and *phaseolicola*. *Phytopathology* **2001**, *91*, 511–518. [CrossRef]
15. Park, S.-W.; Kaimoyo, E.; Kumar, D.; Mosher, S.; Klessig, D.F. Methyl salicylate is a critical mobile signal for plant systemic acquired resistance. *Science* **2007**, *318*, 113–116. [CrossRef]
16. Stintzi, A.; Weber, H.; Reymond, P.; Browse, J.; Farmer, E.E. Plant defense in the absence of jasmonic acid: The role of cyclopentenones. *Proc. Natl. Acad. Sci. USA* **2001**, *98*, 12837–12842. [CrossRef]
17. Bustin, S.A.; Benes, V.; Garson, J.A.; Hellemans, J.; Huggett, J.; Kubista, M.; Mueller, R.; Nolan, T.; Pfaffl, M.W.; Shipley, G.L.; et al. The MIQE guidelines: Minimum information for publication of quantitative real-time PCR experiments. *Clin. Chem.* **2009**, *55*, 611–622. [CrossRef]
18. Schmidt, G.W.; Delaney, S.K. Stable internal reference genes for normalization of real-time RT-PCR in tobacco (*Nicotiana tabacum*) during development and abiotic stress. *Mol. Genet. Genomics.* **2010**, *283*, 233–241. [CrossRef]
19. Kundu, A.; Patel, A.; Pal, A. Defining reference genes for qPCR normalization to study biotic and abiotic stress responses in *Vigna mungo*. *Plant Cell Rep.* **2013**, *32*, 1647–1658. [CrossRef]
20. Wang, Y.J.; Dong, L.; Zhang, C.; Wang, X.Q. Reference gene selection for real-time quantitative PCR normalization in tree peony(*Paeonia suffruticosa* Andr). *J. Agric. Biotechnol.* **2012**, *20*, 521–528. [CrossRef]
21. Yan, H.B.; Ge, W.Y.; Cheng, Y.D.; He, J.G.; Guan, J.F. Cloning and characters analysis of GAPDH gene family in Yali pear. *J. Northwest AF Univ. (Nat. Sci. Ed.).* **2012**, *40*, 181–186. [CrossRef]
22. Chen, Y.; Wu, X.; Gu, C.; Yin, H.; Zhang, S. Selection of reference genes in qRT-PCR of pear "Dangshansuli". *China Fruits* **2018**, *1*, 16–22, 35. [CrossRef]
23. Tang, X.; Zhang, N.; Si, H.J.; Calderón-Urrea, A. Selection and validation of reference genes for RT-qPCR analysis in potato under abiotic stress. *Plant Methods* **2017**, *13*, 85. [CrossRef] [PubMed]
24. Liu, Z.; Qin, Z.L.; Wu, T.; Li, X.M.; Tu, J.F.; Yang, F.C.; Zhu, H.Y. Fluorescence Quantitative Reference Genes and Their Primers and Applications in Leaf Tissues of Two Tree Pears at Different Developmental Stages. CN107988407A. 2018. Available online: https://www.zhangqiaokeyan.com/patent-detail/061204415708.html (accessed on 6 November 2021).
25. Abbas, A.; Yu, H.Y.; Li, X.J.; Cui, H.L.; Chen, J.C.; Huang, P. Selection and validation of reference genes for RT-qPCR analysis in *Aegilops tauschii* (Coss.) under different abiotic stresses. *Int. J. Mol. Sci.* **2021**, *22*, 11017. [CrossRef]
26. Chen, J.Q.; Li, X.Y.; Wang, D.Q.; Li, L.T.; Zhou, H.S.; Liu, Z.; Wu, J.; Wang, P.; Jiang, X.T.; Fabrice, M.R.; et al. Identification and testing of reference genes for gene expression analysis in pollen of *Pyrus bretschneideri*. *Sci. Hortic.* **2015**, *190*, 43–56. [CrossRef]

27. Czechowski, T.; Stitt, M.; Altmann, T.; Udvardi, M.K.; Scheible, W.-R. Genome-wide identification and testing of superior reference genes for transcript normalization in *Arabidopsis*. *Plant Physiol.* **2005**, *139*, 5–17. [CrossRef]
28. Duan, Y.; Tian, C.; Song, Y. In vitro culture and rapid propagation of pear Xuehuali. *Shanxi Agric. Univ. (Nat. Sci. Ed.)*. **2014**, *34*, 464–467. [CrossRef]
29. Jaakola, L.; Pirttilä, A.M.; Halonen, M.; Hohtola, A. Isolation of high quality RNA from Bilberry (*Vaccinium myrtillus* L.) Fruit. *Mol. Biotechnol.* **2001**, *19*, 201–204. [CrossRef]
30. Xie, F.L.; Xiao, P.; Chen, D.L.; Xu, L.; Zhang, B.H. miRDeepFinder:a miRNA analysis tool for deep sequencing of plant small RNAs. *Plant Mol. Biol.* **2012**, *80*, 75–84. [CrossRef]
31. Vandesompele, J.; Preter, K.D.; Pattyn, F.; Poppe, B.; Roy, N.V.; Paepe, A.D.; Speleman, F. Accurate normalization of real-time quantitative RT-PCR data by geometric averaging of multiple internal control genes. *J. Genome Biol.* **2002**, *3*, 34.1–34.11. [CrossRef]
32. Greenberg, J.T.; Yao, N. The role and regulation of programmed cell death in plant-pathogen interactions. *Cell Microbiol.* **2004**, *6*, 201–211. [CrossRef] [PubMed]
33. Sagi, M.; Fluhr, R. Production of reactive oxygen species by plant NADPH oxidases. *Plant Physiol.* **2006**, *141*, 336–340. [CrossRef] [PubMed]
34. Cepauskas, D.; Miliute, I.; Staniene, G.; Gelvonauskiene, D.; Stanys, V.; Jesaitis, A.J.; Baniulis, D. Characterization of apple NADPH oxidase genes and their expression associated with oxidative stress in shoot culture in vitro. *Plant Cell Tissue Organ. Cult.* **2015**, *124*, 621–633. [CrossRef]
35. Cheng, X.; Li, G.H.; Manzoor, M.A.; Wang, H.; Abdullah, M.; Su, X.Q.; Zhang, J.Y.; Jiang, T.S.; Jin, Q.; Cai, Y.P.; et al. In silico genome-wide analysis of respiratory burst oxidase homolog (RBOH) family genes in five fruit-producing trees, and potential functional analysis on lignification of stone cells in Chinese White Pear. *Cells* **2019**, *8*, 520. [CrossRef] [PubMed]
36. Pfaffl, M.W.; Tichopad, A.; Prgomet, C.; Neuvians, T.P. Determination of stable housekeeping genes, differentially regulated target genes and sample integrity: BestKeeper-excel-based tool using pair-wise correlations. *Biotechnol. Lett.* **2004**, *26*, 509–515. [CrossRef] [PubMed]
37. Andersen, C.L.; Jensen, J.L.; Ørntoft, T.F. Normalization of real-time quantitative reverse transcription-PCR data: A model-based variance estimation approach to identify genes suited for normalization, applied to bladder and colon cancer data sets. *Cancer Res.* **2004**, *64*, 5245–5250. [CrossRef]
38. Taylor, S.C.; Nadeau, K.; Abbasi, M.; Lachance, C.; Nguyen, M.; Fenrich, J. The ultimate qPCR experiment: Producing publication quality, reproducible data the first time. *Trends Biotechnol.* **2019**, *37*, 761–774. [CrossRef]
39. Mallona, I.; Lischewski, S.; Weiss, J.; Hause, B.; Egea-Cortines, M. Validation of Reference Genes for Quantitative Real-Time PCR during Leaf and Flower Development in Petunia Hybrida. *BMC Plant Biol.* **2010**, *10*, 4. [CrossRef]
40. Rapacz, M.; Stępień, A.; Skorupa, K. Internal standards for quantitative RT-PCR studies of gene expression under drought treatment in barley (*Hordeum vulgare* L.): The effects of developmental stage and leaf age. *Acta Physiol. Plant.* **2012**, *34*, 1723–1733. [CrossRef]
41. Mafra, V.; Kubo, K.S.; Alves-Ferreira, M.; Ribeiro-Alves, M.; Stuart, R.M.; Boava, L.P.; Rodrigues, C.M.; Machado, M.A. Reference genes for accurate transcript normalization in *Citrus* genotypes under different experimental conditions. *PLoS ONE* **2012**, *7*, e31263. [CrossRef]
42. Xu, Y.Y.; Li, H.; Li, X.G.; Lin, J.; Wang, Z.H.; Yang, Q.S.; Chang, Y.H. Systematic selection and validation of appropriate reference genes for gene expression studies by quantitative real-time PCR in pear. *Acta Physiol. Plant.* **2015**, *37*, 40. [CrossRef]
43. Liu, Z.; Ge, X.X.; Wu, X.M.; Kou, S.J.; Chai, L.J.; Guo, W.W. Selection and validation of suitable reference genes for mRNA qRT-PCR analysis using somatic embryogenic cultures, floral and vegetative tissues in *Citrus*. *Plant Cell Tiss. Organ Cult.* **2013**, *113*, 469–481. [CrossRef]
44. Tong, Z.G.; Gao, Z.H.; Wang, F.; Zhou, J.; Zhang, Z. Selection of reliable reference genes for gene expression studies in peach using real-time PCR. *BMC Mol. Biol.* **2009**, *10*, 71. [CrossRef] [PubMed]
45. Mariot, R.F.; de Oliveira, L.A.; Voorhuijzen, M.M.; Staats, M.; Hutten, R.C.B.; Van Dijk, J.P.; Kok, E.; Frazzon, J. Selection of Reference Genes for Transcriptional Analysis of Edible Tubers of Potato (*Solanum tuberosum* L.). *PLoS ONE* **2015**, *10*, e0120854. [CrossRef]
46. Deng, Y.; Li, Y.D.; Sun, H.Y. Selection of Reference Genes for RT-qPCR Normalization in Blueberry (*Vaccinium corymbosum* × *Angustifolium*) under Various Abiotic Stresses. *FEBS Open Bio* **2020**, *10*, 1418–1435. [CrossRef]
47. Pleskot, R.; Potocký, M.; Pejchar, P.; Linek, J.; Bezvoda, R.; Martinec, J.; Valentová, O.; Novotná, Z.; Žárský, V. Mutual Regulation of Plant Phospholipase D and the Actin Cytoskeleton. *Plant J.* **2010**, *62*, 494–507. [CrossRef]
48. Wang, Y.Z.; Dai, M.S.; Cai, D.Y.; Shi, Z.B. Expression stability analysis of common internal reference genes in pear fruit based on high-throughput sequencing. *Mol. Plant Breed.* **2019**, *17*, 746–753. [CrossRef]

horticulturae

MDPI

Article

The Genes Associated with Jasmonic Acid and Salicylic Acid Are Induced in Tropical Chili Pepper against *Ralstonia solanacearum* by Applying Arbuscular Mycorrhizal Fungi

Erlina Ambarwati [1], Triwidodo Arwiyanto [2], Jaka Widada [3], Taufan Alam [1], Ignatius Putra Andika [4] and Taryono [1,5,*]

1 Department of Agronomy, Faculty of Agriculture, Universitas Gadjah Mada, Jl. Flora, Bulaksumur, Yogyakarta 55281, Indonesia
2 Department of Plant Protection, Faculty of Agriculture, Universitas Gadjah Mada, Jl. Flora, Bulaksumur, Yogyakarta 55281, Indonesia
3 Department of Agricultural Microbiology, Faculty of Agriculture, Universitas Gadjah Mada, Jl. Flora, Bulaksumur, Yogyakarta 55281, Indonesia
4 Faculty of Biotechnology, Universitas Atma Jaya, Jl. Babarsari No. 44, Yogyakarta 55281, Indonesia
5 Agrotechnology Innovation Centre, Universitas Gadjah Mada, Kalitirto, Yogyakarta 55573, Indonesia
* Correspondence: tariono60@ugm.ac.id

Abstract: Controlling soil-borne pathogens is a significant problem in agriculture. Arbuscular mycorrhizae have a potential role in controlling soil-borne pathogens by increasing plant phytohormone contents. However, the mechanism of resistance by mycorrhizae has not been fully elucidated, particularly against bacterial wilt disease in Solanaceae. This study examined the role of mycorrhizae in expressing genes involved in the signaling pathways mediated by jasmonic acid (JA) and salicylic acid (SA) in tropical chili pepper against the bacterium *Ralstonia solanacearum*. Seedlings from ten genotypes of chili pepper were inoculated with a consortium of five mycorrhizal species and/or inoculated with a mixture of nine isolates of *R. solanacearum*. The leaves of 10-week-old plants after the treatment were sampled for real-time polymerase chain reaction analysis. The results showed that the mycorrhizae strengthened the immune system of tropical chili pepper by increasing the relative gene expression levels of JA and SA in genotypes with high and low responsiveness to the mycorrhizae. The relative gene expression level of JA was related to the percentage colonization of mycorrhizae and the resistance of the tropical chili pepper genotypes to *R. solanacearum*. The relative gene expression level of SA was associated with the resistance of tropical chili pepper to *R. solanacearum*.

Keywords: resistance induction; jasmonic acid; salicylic acid; tropical chili pepper; *Ralstonia solanacearum*; arbuscular mycorrhiza

1. Introduction

Pepper and sweet pepper (*Capsicum* spp.) have many uses as spices, natural food coloring agents, and raw materials for making medicines and pesticide mixtures [1–4]. One of the obstacles to chili cultivation during the rainy season is the systemic vascular wilt disease caused by *Ralstonia solanacearum*. *R. solanacearum* is a most devastating soil-borne pathogen, particularly in the Solanaceae family [5–12]. The bacterial wilt pathogen *R. solanacearum* remains difficult to control because of its persistence. This pathogen can survive on plant debris or in the absence of a host. It has a broad host range, high genetic diversity, complex sub-species, a broad geographic distribution, and lives and multiplies in the xylem, thereby blocking the xylem tissue [5,7,13–18].

Current management of pathogens is based on chemical pesticides. However, using chemical pesticides produces residues in crops that are harmful to health, impact the environment, and result in an unsustainable agricultural system [5,19,20]. In recent years, the specific microbiome has been recognized as a potential tool for achieving environmentally friendly agriculture. The importance of the microbiome as a modulator of crop resistance has emerged [5,19,21–24]. The presence of certain microorganisms induces the reprogramming of plant metabolic pathways involved in the defense system and increases the ability of plants to survive under adverse conditions [5,19,23–25].

Endophytic fungi, such as arbuscular mycorrhizae, have potential roles in controlling plant diseases, particularly soil-borne pathogens [26–28]. Local and systemic mycorrhizal-induced resistance (MIR) increases phytohormone contents, such as salicylic acid (SA) and jasmonic acid (JA) [23,24,29–31]. The response of plants to biotrophic pathogens is generally controlled by activating a salicylic acid (SA) dependent response [32]. Beneficial microorganisms modulate the SA pathway [19]. Activating the SA pathway may also influence the favorable performance of the symbiont.

JA is another essential hormone in plant defense that regulates various processes related to plant development, symbiosis, and plant responses to insects and pathogens [32–34]. JA stimulates induced systemic resistance (ISR) activated by the interactions between plants and certain beneficial microorganisms. Antagonism of the JA and SA biosynthetic pathways occurs [35]; however, some researchers have reported synergistic interactions between them [36,37]. Modulation of the JA signaling pathway may be involved in the symbiosis by keeping the endophytic fungi in the asymptomatic stage [31,38].

Managing beneficial microbes for plants has been proposed as a new platform to revolutionize plant protection against pathogens. However, the mechanisms underlying the symbiotic relationship between plants and beneficial microorganisms are still being investigated. Several studies have shown that biocontrol agents efficiently reduce the incidence of bacterial wilt disease in greenhouses, but field trials have not yielded results [5].

The present study examined the symbiotic relationship between tropical chili pepper genotypes and mycorrhizae. Genotypes of peppers with different responsiveness to mycorrhizae and different resistance to bacterial wilt disease were studied. This study aimed to investigate the potential of mycorrhizae to increase the resistance of the tropical chili pepper genotypes against *R. solanacearum* through hormone signaling pathways. The potential of mycorrhizae was studied in relation to the relative expression of the genes associated with JA and SA signaling pathways. The effect of mycorrhizae on phenotypic traits of the tropical chili pepper was also investigated.

2. Materials and Methods

2.1. Chili Pepper Genotypes, Mycorrhizae, and the Ralstonia solanacearum Isolates

This study was carried out at the Agrotechnology Innovation Center of the Universitas Gadjah Mada at Kalitirto Yogyakarta, Indonesia (7°79′58.59″ N, 110°46′52.85″ E). The experiment examined the relative gene expression of JA and SA at the generative growth phase when the plant started flowering, of ten tropical chili pepper genotypes with different responses to mycorrhizae and resistance to *R. solanacearum*.

Ten chili pepper genotypes were used in this study. The genotypes consisted of eight accessions from the Genebank of the Universitas Gadjah Mada and East-West Seed Indonesia (Ewindo) and two commercial cultivars. Based on the published descriptions of the cultivars, the two commercial cultivars were used as the resistant check cultivar (C-50, Known You Seed) and the susceptible check cultivar (C-54, Oriental Seed) against *R. solanacearum*. The mycorrhizal inoculum was a mixture of *Glomus* sp., *Funneliformis* sp., *Acaulospora* sp., *Gigaspora* sp., and *Scutellospora* sp. with a zeolite carrier obtained from the Agricultural Microbiology Laboratory, Faculty of Agriculture, Universitas Gadjah Mada. Mycorrhizal content was 500 inoculums per gram of zeolite carrier. The *R. solanacearum* isolate used was a mixture of nine bacterial isolates.

2.2. Procedure of Experiments

The study consisted of three experiments in parallel using separate sets of experimental plants.

Experiment 1a. *Evaluation of genotypes responsiveness to mycorrhizae*

The seeds of the ten chili pepper genotypes were germinated in a germination box for 5 days. Twenty seedlings of each genotype were transplanted to nursery pot trays filled with sterilized growing media. The medium was a mixture of soil, coco peat, and husk charcoal at a ratio of 2:1:1 (*v:v:v*). Before transplanting the seedlings, the medium was mixed with the mycorrhizal inoculum at a ratio of 10:1 (*w:w*). The seedlings were maintained in the greenhouse until they were 21 days old.

The 21-day-old seedlings were transplanted into a 10-L planter bag filled with 7 kg of planting media per planter bag. The planting media was a mixture of soil, compost, coco peat, and husk charcoal at a ratio of 2:1:1:1 (*v:v:v:v*). The plants were maintained in the screen house until 10 weeks old.

The growth response to mycorrhizal symbiosis was defined as an increase in plant biomass inoculated with mycorrhiza compared to the biomass of plants that were not inoculated [39]. The plants were sampled 10 weeks after transplant to examine the dry biomass. The plant biomass was represented by shoot and root dry weight. Dry shoots and roots were weighed after the biomass was sun-dried in a greenhouse for seven days and oven-baked at 68 °C for 48 h. The percentage of mycorrhizal colonization was determined by staining the roots using trypan blue 0.05% (*w:v*) as described in the previous study [40] with modifications. The percentage of mycorrhizal colonization was calculated from the proportion of the mycorrhizal root area to the total root area. The area was measured using the Fiji program [41].

Experiment 1b. *Evaluation of genotypes resistance to R. solanacearum*

The seeding procedure was conducted similarly to Experiment 1a. Shortly after the seedlings were transplanted into the nursery pot trays, the *R. solanacearum* was inoculated using the drenching method. A mixture of nine strains of the *R. solanacearum* bacterial suspension at a concentration of 1×10^8 CFU mL^{-1} (OD600 = 0.1) was poured (20 mL of each) into the planting medium around the root area.

The resistance of the ten tropical chili pepper genotypes to *R. solanacearum* was evaluated based on the wilt symptoms of seedlings The wilt symptoms in the seedlings were observed in the morning on days 3, 7, 14, 21, 28, and 35 after the bacterial inoculation based on the scoring method as described in the previous study [7,17] with modifications. Score 0: all leaves or plants remained healthy with no symptoms; score 1: $\leq$20% withered leaves or plants; score 2: 21–40% withered leaves or plants; score 3: 41–60% withered leaves or plants; score 4: 61–80% of withered leaves or plants; and score 5: >80% withered leaves or plants/dead plant. Disease intensity (DI) and disease incidence (DInc.), and the area under the disease progress curve (AUDPC) were calculated using the formula as referred to from the previous study [17,42,43], respectively. The DI was used to determine the resistance of each genotype, including the check cultivar. Genotypes were categorized as resistant if the DI was <30%, moderately resistant if the DI was 31–40%, moderately susceptible if the DI was 41–50%, and susceptible if the DI was >50% [17].

Experiment 2. *Evaluation of relative gene expression JA and SA*

The second experiment combined the responsiveness of the genotypes to mycorrhizae with their resistance to *R. solanacearum* to evaluate the relative gene expression of JA and SA. The treatments applied in this study consisted of seedlings that were not inoculated with mycorrhizae and not inoculated with *R. solanacearum* (M0R0), seedlings that were inoculated with mycorrhiza but not inoculated with *R. solanacearum* (M1R0), seedlings that were not inoculated with mycorrhiza but inoculated with *R. solanacearum* (M0R1), and seedlings that were inoculated with mycorrhiza and inoculated with *R. solanacearum* (M1R1).

The seeds of the ten chili pepper genotypes were germinated in a germination box for 5 days. Twenty seedlings of each genotype were transplanted to nursery pot trays filled with sterilized growing media. The medium was a mixture of soil, coco peat, and husk charcoal at a ratio of 2:1:1 (*v:v:v*). Before transplanting the seedlings, the medium was mixed with the mycorrhizal inoculum at a ratio of 10:1 (*w:w*). The seedlings were maintained in the greenhouse until they were 21 days old.

The 21-day-old seedlings were transplanted into a 10-L planter bag filled with 7 kg of planting media per planter bag. The planting media was a mixture of soil, compost, coco peat, and husk charcoal at a ratio of 2:1:1:1 (*v:v:v:v*). The bacterium was inoculated on 21-day-old seedlings using the soil drenching method. A mixture of nine strains of the *R. solanacearum* suspension at a concentration of 1×10^8 CFU mL^{-1} (OD600 = 0.1) was poured into the seedling media around the root area. Then, the seedlings were watered in the morning and afternoon, as required. A total of 20 planter bags were prepared for each genotype and repeated three times for each treatment. The bags were arranged in a completely randomized block design. The seedlings inoculated with *R. solanacearum* were placed in separate screen houses from those that were not inoculated. The plants were raised until they entered the generative growth phase, and leaf samples were taken (approximately 10 weeks after transplanting).

2.3. Isolation of Leaf RNA

The expression of JA and SA pathway genes was detected by quantitative real-time reverse transcription-polymerase chain reaction (qRT-PCR) (BIO-RAD CFX96tm Real-Time System; Bio-Rad Laboratories, Hercules, CA, USA) using RNA samples extracted from leaf tissue. RNA was isolated from the youngest leaf tissue for qRT-PCR using the Geneaid RNA mini kit (Plant) [44]. A sample of 100 mg of leaf tissue was crushed in liquid nitrogen. The crushed tissue was placed in a 1.5 mL microtube and 500 µL of PRB Buffer and 5 µL of β-mercaptoethanol was added and mixed until homogeneous. The mixture was incubated at 60 °C for 5 min, transferred to a filter column with a 2 mL collection tube, and centrifuged at 3000 rpm (1000× *g*) for 1 min. Then, the filter column was discarded. The supernatant was taken from the collection tube and transferred to a 1.5 mL microtube. Absolute ethanol was added to the supernatant at half the volume of the total supernatant taken and homogenized with a vortex mixer. The supernatant was transferred to an RB column with a 2 mL collection tube and centrifuged at 12,000 rpm (16,000× *g*) for 1 min. The 2 mL of solution in the collection tube was discarded, the supernatant was returned to the RB column, and 400 µL of W1 buffer was added and centrifuged at 12,000 rpm (16,000× *g*) for 30 s. The solution in the collection tube was discarded and the supernatant was returned to the RB column. Then, 600 µL of wash buffer was added to the RB column, the eluate was centrifuged at 12,000 rpm (16,000× *g*) for 1 min, and the solution in the collection tube was discarded. The washing step with the wash buffer was repeated. The column tube was dried by centrifugation at 12,000 rpm (16,000× *g*) for 3 min. Then, the RB column was transferred to a 1.5 mL microtube, and 50 µL of RNAse-free water was added to the center of the RB column and allowed to stand for 2 min. In the last stage, the eluate was centrifuged at 12,000 rpm (16,000× *g*) for 1 min, and the RNA was eluted into a 1.5 mL microtube.

2.4. qRT-PCR for GAPDH-cp and Genes of Interest

The qRT-PCR analysis was performed using the Bio-Rad CFX-96 real-time PCR system. The target genes were oxophytodienoic acid reductase-3 (OPR3) (*jasmonic acid*), and ICL (*salicylic acid*), and the reference gene was GAPDH-cp (Table 1). The total reaction volume was 20 µL, including 10 µL of 2 SensiFAST SYBR No-ROX One-Step Mix, 0.2 µL of reverse transcriptase, 0.4 µL of riboSafe RNAse inhibitor, 1.6 µL of the primers (0.8 µL of the10 µM forward primer and 0.8 µL of the 10 µM reverse primer), 4 µL of the RNA template, and

3.8 μL of RNAse-free water. The qRT-PCR was implemented in three stages. Stage I was reverse transcription at 45 °C for 10 min, stage II was polymerase activation at 95 °C for 2 min, and stage III was 35 cycles of denaturation at 95 °C for 1 min, annealing for 30 s at the optimum temperature for each primer, elongation at 72 °C for 1 min, and a melting curve analysis at 65–95 °C for 5 s. The analysis was repeated in triplicate for each gene. The relative quantification of the genes of interest was assessed using the $2^{-\Delta\Delta CT}$ method [45].

Table 1. Primer and reference genes used in the real-time RT-PCR analysis.

Target	Primer ID	Sequence (5′–3′)	Ta (°C)	References
Jasmonic acid	OPR3 Forward	CTTCTCATGCAGTGTATCAAC	46.4	[46]
	OPR3 Reverse	CGTCCAAGTGATCTATAGCTG		
Salicylic acid	ICL Forward	GCGTCGGAGGAAGTCAAGAA	50.5	[47]
	ICL Reverse	GGATTGGAACTTGGAGCCGA		
Reference gene	GAPDH-Cp Forward	ATGATGATGTGAAAGCAGCG	46.4/50.5	-
	GAPDH-Cp Reverse	TTTCAACTGGTGGCTGCTAC		

2.5. Phenotypic Response of the Genotypes to Mycorrhizae and *R. solanacearum*

The phenotypic response was analyzed from the data of plant height and stem diameter of ten tropical chili pepper genotypes from all the treatments, i.e., M0R0, M1R0, M0R1 and M1R1. Plant height and stem diameter were measured at 10 weeks after transplanting. Plant height was measured from the ground surface to the shoot tip. The stem diameter was measured at approximately 5 cm above the ground. The data was collected from 15 plants for each genotype.

2.6. Data Analysis

The statistical analysis was performed using R Studio software version 3.6.2 [48]. The means of all observed traits were calculated using the least square means and analysis of variance. The Scott-Knott post hoc analysis was used to detect differences between the mean values of the observed characteristics. A p-value < 0.05 was considered significant. Structural equation modeling-partial least square (SEM-PLS) was used to construct the path coefficient and assess the cumulative effect of genotype responsiveness to mycorrhizae and the genotype resistance to *R. solanacearum* on the relative expression of the JA and SA genes [49,50]. A standardized stepwise regression was used to investigate the parameters of each genotype responsiveness variable to the mycorrhizae and the resistance of the chili genotypes to *R. solanacearum*, which affected the relative expression of the JA and SA genes [51]. SEM-PLS analysis and standardized stepwise regression were performed using PROC GLM and PROC REG in SAS 9.4 [52]. A heatmap was created using the Euclidean distance and the average heatmap method in R [53]. The heatmap was used to classify the genotypes based on their responsiveness to the mycorrhizae, resistance to *R. solanacearum*, and the relative gene expression of JA and SA. The heatmap was also used to determine the responsiveness to the mycorrhizae and/or the resistance to *R. solanacearum* related to the relative gene expression of JA and SA.

3. Results

3.1. Evaluation of the Growth Response to Arbuscular Mycorrhizae and Resistance to *R. solanacearum*

The growth response by the ten genotypes to the mycorrhizae was divided into three groups. Genotype C-38 had the highest growth response to the mycorrhizae, reaching 388.00% ± 37.40%, while genotype C-37 had the lowest growth response of 27.00% ± 2.60%. The growth response to mycorrhizae of four genotypes (C-08, C-30, C-09, C-38) was high; the responses of the other four genotypes (C-34, C-49, C-02, C-37) and the two commercial cultivars (C-50, C-54) were low (Figure 1).

All tropical chili pepper genotypes were infected with *R. solanacearum* 35 days after inoculation (DAI) but with different DI values. Among the eight accessions tested, three were classified as resistant, one was moderately resistant, and four were susceptible genotypes (Figure 1). The susceptible check had a DI value of 67.19% ± 0.19%. The four accessions susceptible to *R. solanacearum* had DIs of 51.97–77.61%. The susceptible check in this evaluation resulted in DI, which was the same as the cultivar description. The resistant check cultivar had a DI of 16.22% ± 0.07% at 35 DAI, which was the same as the cultivar description. Three accessions classified as resistant had higher DIs than the resistant check, ranging from 44.52 to 59.21% and 20.96 to 23.22%, respectively. One accession classified as moderately resistant had a DI of 31.91% ± 0.04% (Figure 1).

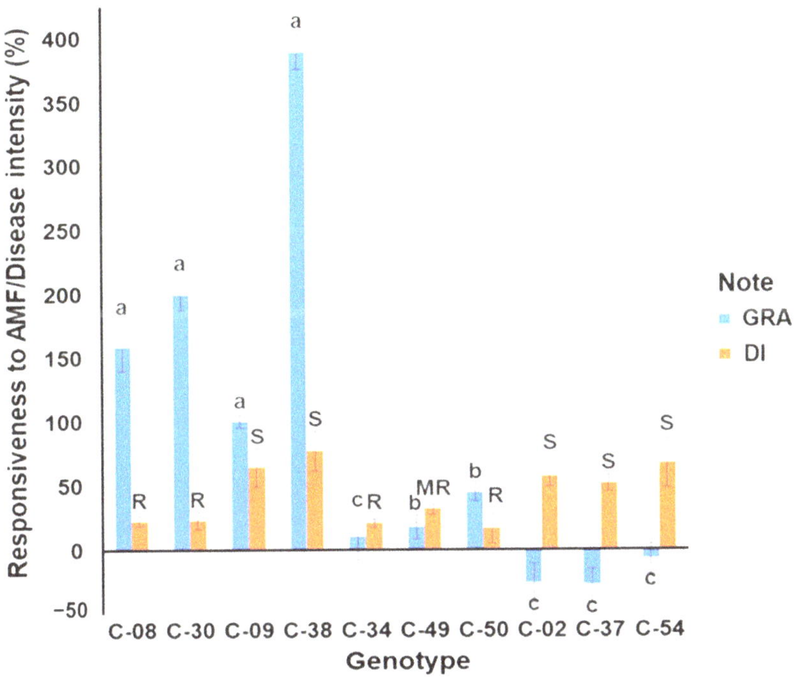

Figure 1. The growth responsiveness of the tropical chili pepper genotypes to mycorrhizae (GRA) and the disease intensity of the tropical chili pepper genotypes to *R. solanacearum* (DI) at 35 days after inoculation (DAI). Note: The same lowercase letters in the GRA histogram indicate no significant differences between the genotypes at a significance level of 0.95 according to the Scott-Knott test. The uppercase letters in the DI histogram indicate the resistance of the genotypes based on the disease intensity at 35 DAI. Bars represent the mean ± standard deviation. AMF: arbuscular mycorrhizal fungi. GRA: the growth response of genotype to mycorrhiza; DI: disease intensity at 35 DAI. R: resistant; MR: moderately resistant; and S: susceptible; C-50: resistant check cultivar and C-54: susceptible check cultivar to *R. solanacearum* according to their published cultivar's description.

3.2. Effect of Arbuscular Mycorrhizae on the Relative Gene Expression of JA

The relative gene expression of JA in genotypes with high responsiveness to mycorrhizae and resistance to *R. solanacearum* (C-08 and C-30) was highest in the M1R1 treatment. In genotypes with high responsiveness to mycorrhizae and susceptibility to *R. solanacearum* (C-09 and C-38), the relative gene expression of JA was highest in M0R1. In genotypes with low responsiveness to mycorrhizae and susceptibility to *R. solanacearum* (C-02 and C-37), the relative gene expression of JA was highest in the M0R1 treatment. In genotypes with low responsiveness to mycorrhizae and resistance or moderate resistance to *R. solanacearum* (C-34 and C-49), the relative gene expression of JA was highest in M1R1 (Figure 2).

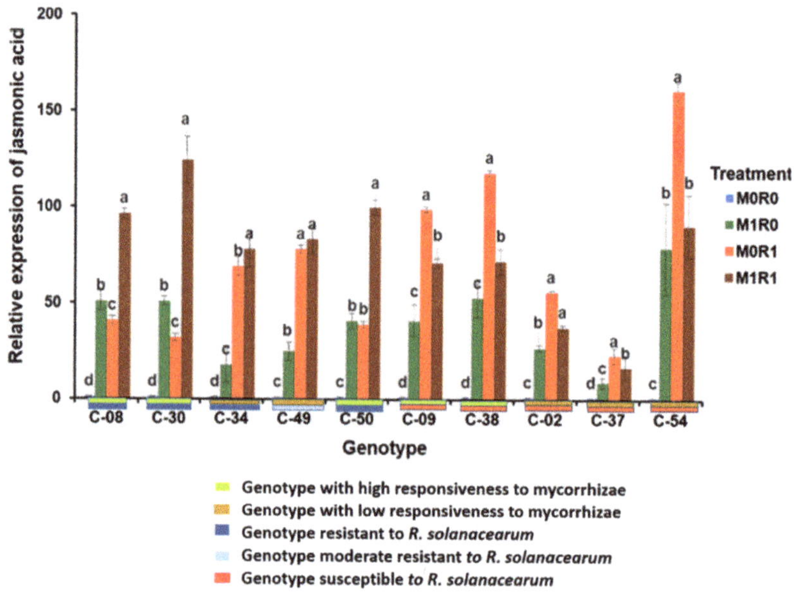

Figure 2. Relative gene expression of jasmonic acid (JA) in the tropical chili pepper genotypes. Note: Plants without mycorrhizae (M0); plants inoculated with mycorrhizae (M1); plants without *R. solanacearum* (R0); plants inoculated with *R. solanacearum* (R1). Bars represent the mean ± standard deviation. The same lowercase letters indicate no significant difference in the mean values between treatments in each genotype, according to the Scott-Knott test, at a significance level of 0.95.

Inoculating mycorrhizae but not *R. solanacearum* (M1R0) showed that mycorrhizae caused the relative gene expression of JA to increase in the chili pepper genotypes. The relative gene expression of JA was higher in genotypes with high responsiveness to mycorrhizae than in those with low responsiveness. The relative gene expression of JA in the genotypes with high responsiveness to mycorrhizae ranged from 40.65 to 53.31 times, while the genotypes with low responsiveness only ranged from 9.33 to 27.20 times.

Inoculating with *R. solanacearum* but not the mycorrhizae (M0R1) showed that the relative gene expression of JA from the genotypes resistant to *R. solanacearum* was lower than that of the susceptible genotypes. However, the opposite result occurred in the genotypes resistant to *R. solanacearum*. The relative gene expression of JA from genotypes resistant to *R. solanacearum* ranged from 31.64 to 78.14. The relative gene expression of JA in the genotypes susceptible to *R. solanacearum* went from 38.38 to 118.04 times.

The genotypes inoculated with *R. solanacearum* and mycorrhizae (M1R1) triggered the relative gene expression of JA. The relative magnitude of JA gene expression depended on the resistance of the genotype to *R. solanacearum* and the responsiveness to the mycorrhizae. The relative gene expression of JA in the genotypes resistant to *R. solanacearum* increased in the range of 77.88–124.44 times Genotypes susceptible to *R. solanacearum* in the presence of mycorrhizal symbiosis increased their relative gene expression of JA in the range of 16.88–71.71 times.

3.3. Effect of Arbuscular Mycorrhizae on the Relative Gene Expression of SA

The relative gene expression of SA in the genotypes with high responsiveness to mycorrhizae and resistance to *R. solanacearum* (M0R1) (C-08 and C-30) was the highest. The relative SA gene expression in the M1R1 and M1R0 treatments was higher than in the M0R1 in the resistant check cultivar (C-50). In the genotypes with high responsiveness to mycorrhizae and susceptibility to *R. solanacearum* (C-09 and C-38), the relative gene expression of SA in the M1R1 treatment was highest. The relative gene expression of SA in the resistant check cultivar was highest in the M1R1 treatment. In the genotypes with low responsiveness to mycorrhizae and susceptibility to *R. solanacearum* (C-02 and C-37), the relative gene expression of SA was highest in the M1R1 treatment. In the genotypes with low responsiveness to mycorrhizae and resistance or moderate resistance to *R. solanacearum* (C-34 and C-49), the relative gene expression of SA was highest in the M1R1 treatment, followed by M1R0 and M0R1 (Figure 3).

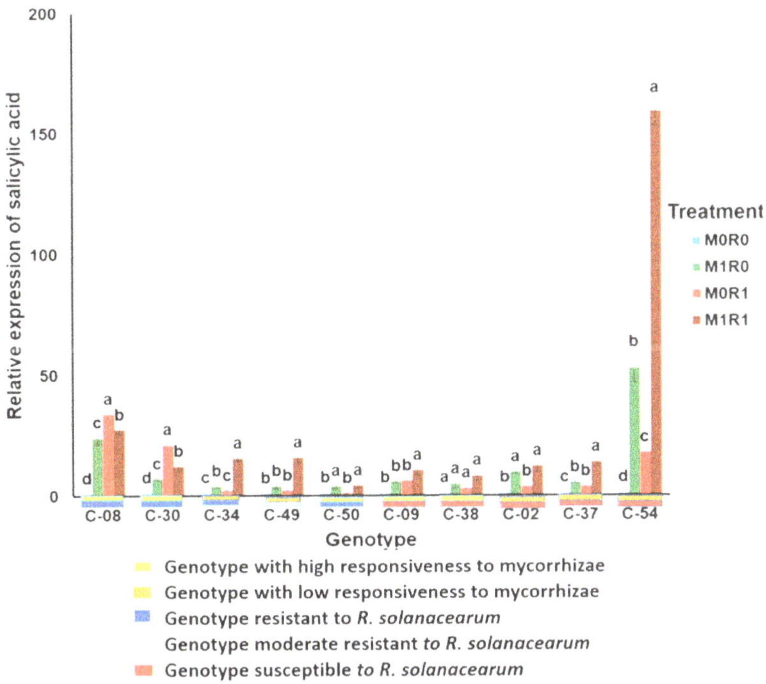

Figure 3. Relative gene expression of salicylic acid (SA) in the tropical chili pepper genotypes. Note: Plants without mycorrhizae (M0); plants inoculated with mycorrhizae (M1); plants not inoculated with *R. solanacearum* (R0); plants inoculated with *R. solanacearum* (R1). Bars represent mean ± standard deviation. The same lowercase letters indicate no significant difference in mean values between treatments in each genotype according to the Scott-Knott test at a significance level of 0.95.

Genotypes inoculated with mycorrhizae but not with *R. solanacearum* (M1R0) showed the relative gene expression of SA depended on the responsiveness of the genotype to the arbuscular mycorrhizal. Genotypes with high responsiveness to the mycorrhizae revealed higher relative gene expression of SA than genotypes with low responsiveness. The range of relative gene expression of SA in the responsive genotypes was 4.81–23.7 times, which was higher than those in genotypes with low responsiveness at 4.15–9.63 times.

The resistant genotypes tended to express more SA than the susceptible genotypes. This observation can be seen in the genotypes treated with bacterial inoculation without mycorrhizae (M0R1). The relative gene expression range of SA in the resistant genotypes was 2.93–34.21 times, which was higher than that in the susceptible genotypes at 2.02–17.68 times.

The combined inoculation treatment with *R. solanacearum* and mycorrhizae (M1R1) showed that the arbuscular mycorrhizal symbiosis increased the relative gene expression of SA. The increase in relative SA expression depended on the resistance of the genotype to *R. solanacearum*. In the presence of mycorrhizae, the relative gene expression of SA in the resistant genotypes ranged from 12.16 to 74.45 times. It tended to be higher than that in the susceptible genotypes which were only 8.13–13.81 times.

3.4. Structural Equation Modeling-PLS and Standardized Stepwise Regression

The results of the SEM-PLS analysis showed that the resistance to *R. solanacearum* had a significant path coefficient on the relative gene expression of JA ($p < 0.048^*$) (Figure 4). However, the relative gene expression of SA was not directly affected by either the responsiveness to mycorrhizae or the resistance to *R. solanacearum* (Figure 5).

The standardized stepwise regression showed that the relative gene expression of JA was significantly affected by the percentage colonization of mycorrhizae (PCA) (0.935***) ($R^2 = 0.874^{***}$) and the DInc (0.9480***) ($R^2 = 0.928^{***}$). Similarly, the relative expression of the SA gene was significantly affected only by DInc (0.648*) ($R^2 = 0.648^*$). Thus, the PCA is a responsiveness variable, and DInc is a variable of resistance. Therefore, the relative gene expression of JA was strongly affected by the responsiveness of the tropical chili pepper genotypes to mycorrhizae and the resistance to *R. solanacearum*. The relative gene expression of SA was related to the resistance of the tropical chili pepper genotypes to *R. solanacearum*.

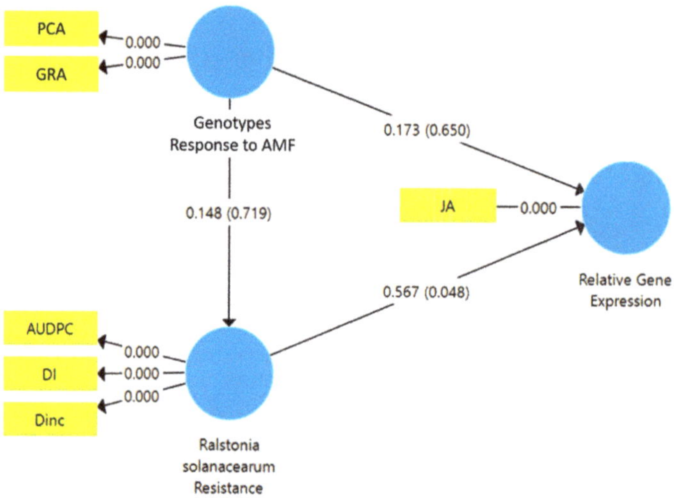

Figure 4. Structural equation modeling-partial least square (SEM-PLS) of the relationship between the responsiveness to mycorrhizae and the resistance of the tropical chili pepper genotypes to the *R. solanacearum* variables and the relative gene expression of JA. AMF: arbuscular mycorrhizal fungi. PCA: percentage colonization of mycorrhiza, GRA: the growth response of the tropical chili pepper genotype to mycorrhizae, DI: disease intensity, DInc.: disease incidence, AUDPC: area under disease progress curve, JA: jasmonic acid.

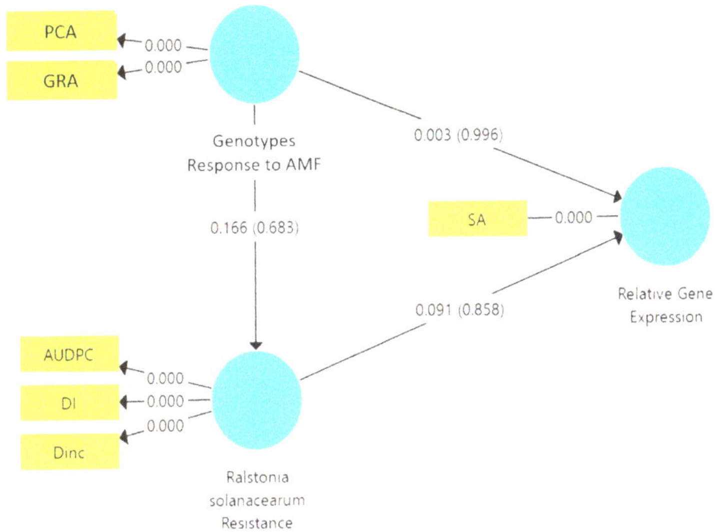

Figure 5. Structural equation modeling-partial least square (SEM-PLS) of the relationship between the responsiveness to mycorrhizae and the resistance of the tropical chili pepper genotypes to *R. solanacearum* variables and the relative gene expression of SA. AMF: arbuscular mycorrhizal fungi. PCA: percentage colonization of mycorrhizae, GRA: the growth response of the tropical chili pepper genotypes to mycorrhizae, DI: disease intensity, DInc.: disease incidence, AUDPC: area under disease progress curve, SA: salicylic acid.

3.5. Clustering of the Genotypes Based on the Responsiveness to Mycorrhizae, Resistance to R. solanacearum, and the Relative Gene Expression of JA and SA

The genotypes were grouped into four clusters based on the heatmap (Figure 6). The heatmap shows high diversity in several traits, i.e., PCA, genotype responsiveness to mycorrhizae (GRA), DI, DInc, the AUDPC, and the relative gene expression of JA and SA. The relative variation in JA and SA gene expression in the tropical chili pepper genotypes was similar (Figures 2 and 3). The heatmap shows that the relative gene expression of SA was related to variable resistance of the genotypes to *R. solanacearum*. In contrast, the relative gene expression of JA was related to variable resistance of the genotypes to *R. solanacearum* and the GRA. The results of this heatmap agree with the standardized stepwise regression results regarding the relative determinants of JA and SA gene expression.

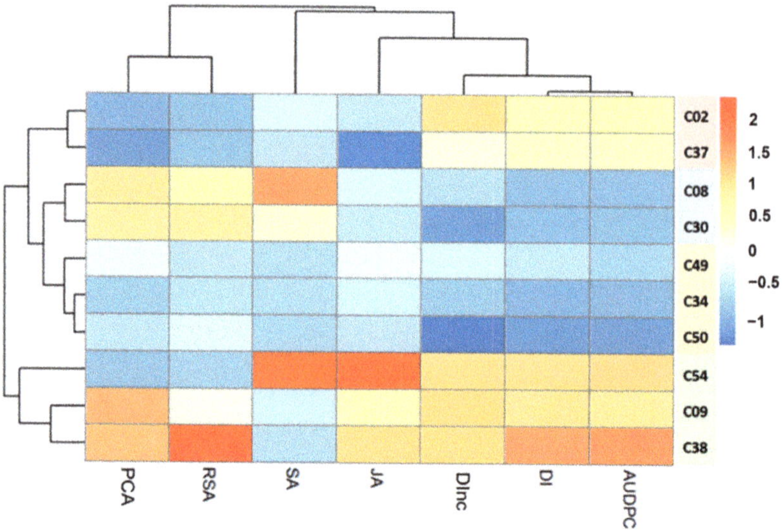

Figure 6. Heatmap describing the diversity of responsiveness to the mycorrhizae, the resistance to *R. solanacearum*, and the relative gene expression of JA and SA. Blue and red indicate an increase and decrease in the value of each trait, respectively. The black lines on the top and at the left of the heatmap indicate dendogram of variables and genotypes, respectively. Color-bar on the right side of the dendogram indicates the genotypes in the same cluster. PCA: percentage colonization of mycorrhiza, GRA: genotype responsiveness to mycorrhiza, SA: salicylic acid, JA: jasmonic acid, DInc: disease incidence, DI: disease intensity, AUDPC: area under the disease progress curve.

3.6. Phenotypic Response Based on the Responsiveness to Mycorrhizae and Resistance to R. solanacearum

Figure 7 shows the response of ten genotypes of tropical chili pepper based on responsiveness to mycorrhizae, resistance to *R. solanacearum*, and relative gene expression of JA and SA on plant height. In general, the genotypes inoculated with *R. solanacearum* and arbuscular mycorrhizae (M1R1) were able to suppress the growth of *R. solanacearum*, the plants were taller than plants infected *R. solanacearum* but not inoculation mycorrhizae (M0R1). The same phenomenon occurs in the variable stem diameters of ten genotypes with various mycorrhizal responsiveness and *R. solanacearum* resistance, mycorrhizae were able to suppress the growth of *R. solanacearum* (Figure 8). The relative gene expression of JA and SA induced by mycorrhizae in the presence of *R. solanacearum* (M0R1) could suppress the growth of *R. solanacearum*. This was indicated by the higher plants and larger stem diameters of the tropical chili pepper genotypes than the plants infected *R. solanacearum* but not inoculation mycorrhizae (M0R1) (Figures 7 and 8).

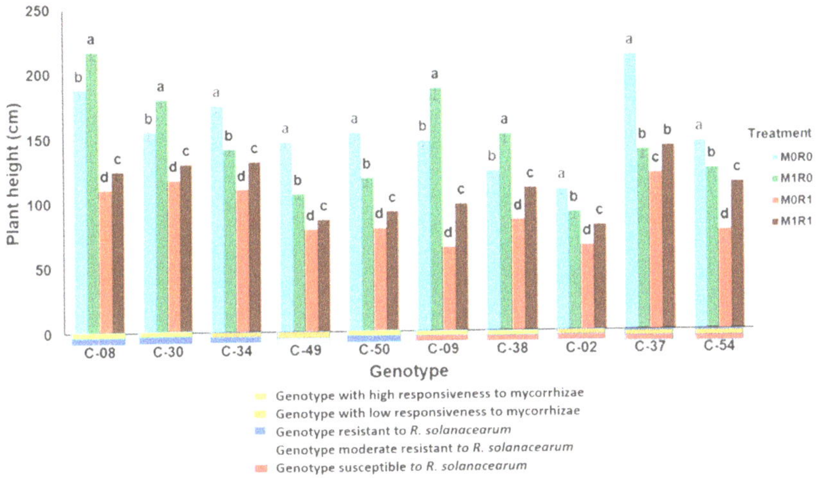

Figure 7. Plant height of ten genotypes tropical chili pepper as a phenotypic response to the treatments. Bars represent mean ± standard deviation. The same lowercase letters indicate no significant difference in mean values between treatments in each genotype according to the Scott-Knott test at a significance level of 0.95.

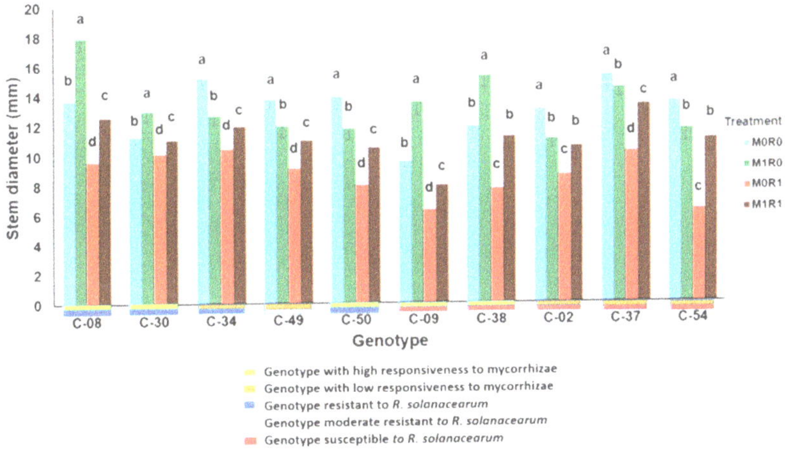

Figure 8. Stem diameter of ten genotypes tropical chili pepper as a phenotypic response to the treatments. Bars represent mean ± standard deviation. The same lowercase letters indicate no significant difference in mean values between treatments in each genotype according to the Scott-Knott test at a significance level of 0.95.

4. Discussion

4.1. Responsiveness to Arbuscular Mycorrhizae and the Resistance to R. solanacearum

Figure 1 shows the diverse growth responses of the tropical chili pepper genotypes to the mycorrhizae. The genotypes differed in their response to arbuscular mycorrhizal colonization. The effectiveness of the symbiosis with mycorrhizae is highly dependent on the species and the genotype of the host plant [54,55]. The genotype is an essential factor determining the response to mycorrhizae [31,56,57]. Differences in the response to mycorrhizae emphasize a genetic basis for the interactions between plants and mycorrhizae [31]. Many studies have reported significant differences in the responsiveness to mycorrhizae between genotypes within the same host plant species [27,54,57]. The symbiosis with arbuscular mycorrhizae changed the growth of the tropical chili peppers. Janos [58] stated that the responses of plant genotypes to mycorrhizal colonization can be classified into positive, neutral, and negative. The positive reactions of four of the tropical chili pepper genotypes indicated that mycorrhizal colonies and the host plant got benefit from the symbiosis. A small positive symbiotic response indicated that the tropical chili pepper as the host plant did not respond to the presence of mycorrhizae, and the mycorrhizae did not interfere with the host plant. This observation indicates that there was no effect on the host plant in the presence of arbuscular mycorrhizal symbioses.

In contrast, a negative response was found in two of the tropical chili pepper genotypes, as the mycorrhizal symbioses inhibited plant growth or induced symbiotic parasitism. The reaction to the mycorrhizae is neutral or negative in cases where mycorrhizal colonization does not increase plant growth [27]; the biomass produced by plants does not increase even though the plants are well colonized with mycorrhizae [57].

Screening for resistance to *R. solanacearum* in chilies has been carried out but has focused on red chilies and sweet pepper (*Capsicum annuum*). Several sources of resistance to bacterial wilt disease have been identified in *Capsicum* pepper [12,42,59–64]. Three chili pepper genotypes from the germplasm were resistant, one was moderately resistant, and four were susceptible (Figure 1). These results demonstrate that resistant genotypes can be obtained from Indonesia, as reported by several researchers [62,64,65]. This also explains why Southeast Asia is an essential source of chili germplasm resistant to *R. solanacearum* bacterial wilt disease. The germplasms were more adapted to the various environment. They collected represented centers of chili pepper production areas in the three provinces in Jawa Island. Resistant genotypes are used as a source of resistance genes in breeding programs. As *R. solanacearum* easily adapts to the environment, including infecting resistant host plants, the screening results in this study could be used to breed new resistant cultivars.

4.2. Interaction between the Genotypes of Tropical Chili Pepper, the Arbuscular Mycorrhizae, and R. solanacearum on the Relative Gene Expression of JA and SA, and Phenotypic Response

Arbuscular mycorrhizal fungi change the community and the biotic interactions in the rhizosphere. This mycorhizospheric effect alters the communities of beneficial microorganisms and suppresses pathogens in the soil [66–68]. Associations with mycorrhizae directly benefit plants by producing secondary metabolites or modulating plant defenses. A recent study showed that different endophytes induce plant resistance to biotic or abiotic stressors [32,69]. However, the plant microbiome is strongly influenced by its host. Plants must recognize endophytes during colonization, as signaling molecules, and identifying microbes as beneficial partners will create a symbiotic mutualism.

All of these factors affect different plant species in different microbiomes [19]. This study investigated the function of the symbiosis of chili pepper and mycorrhizae and the resistance to *R. solanacearum.* Ten tropical chili pepper genotypes with different responsiveness to mycorrhizae and resistance to *R. solanacearum* were tested. The function of the mycorrhizae was evaluated and the relative gene expression of JA and SA in each tropical chili pepper genotype was determined.

JA is an essential hormone involved in the plant resistance system. JA regulates various processes related to plant development, symbiotic interactions, and plant responses to pathogens, particularly soil-borne pathogens. JA triggers ISR, which is activated by the relationship between the plant and certain beneficial microorganisms. ISR is marked by the accumulation of JA [19,32,33]. The relative gene expression of JA in the ten tropical chili pepper genotypes is shown in Figures 2 and 6. In all ten tropical chili pepper genotypes, treatment with mycorrhizae and *R. solanacearum* infection (M1R1) increased the relative gene expression of JA. The increase in the relative gene expression of JA varied (Figure 2). For the resistant and moderately resistant genotypes (with blue bar under the X-axis), the relative gene expression of JA was higher in the M1R1 treatments. In this genotype group, the difference between M1R1 and M0R1 was larger in the genotypes which are responsive to mycorrhizae (with green and blue bar under the X-axis). It means that the effect of mycorrhizae was higher in the genotypes with high responsiveness to mycorrhizae. In contrast, the relative gene expression of JA in the five susceptible genotypes was higher in the M0R1 treatment (orange histograms with red bar under the X-axis). This finding indicates that JA is always expressed in the infected plants, but with different expressions between resistant and susceptible plants. For resistant plants, the relative expression of JA was depended on responsiveness to mycorrhiza. Genotypes C-08, C-30, and C-50 were resistant to *R. solanacearum* and had a high responsiveness to mycorrhizae. It showed a high expression of JA even in the treatment without infection. Meanwhile, for the susceptible plants, the expression of JA depended mainly on the presence *R. solanacearum* infection.

R. solanacearum infection increased the expression of the OPR3 gene, and the relative gene expression of JA was higher (Figure 2). The SEM-PLS in Figure 4 shows a relationship between JA gene expression and the resistance of the genotypes to *R. solanacearum*. The heatmap revealed similar results (Figure 6). As the accumulation JA was altered by inoculating the endophytic fungi, it can be argued that the expression of several marker genes for different steps in the JA biosynthetic pathway and other genes is associated with plant resistance [19].

The M1R0 treatment increased the relative gene expression of JA in all ten tropical chili pepper genotypes. However, differences were observed in the relative gene expression of JA. The relative gene expression of JA was higher in genotypes with a high mycorrhizal response than in those inoculated with *R. solanacearum* only (M0R1). A similar result was found for the resistant check cultivar. However, this result differed from genotypes with high responsiveness to the mycorrhizae but susceptibility to *R. solanacearum* in which the relative gene expression of JA was the lowest. This result is the same as in genotypes classified as poor responders to arbuscular mycorrhizae. A similar result was found in the susceptible check cultivar. The JA signaling involved in symbiosis must be modulated to maintain arbuscular mycorrhizal fungal endophytes in the asymptomatic stage [19,38].

The relative gene expression of JA in the ten genotypes inoculated with arbuscular mycorrhizae and *R. solanacearum* (M1R1) increased, even in the genotypes with a low or negative response to the mycorrhizae. The mycorrhizae increased the relative gene expression of JA in all ten tropical chili pepper genotypes that were inoculated with *R. solanacearum*. Based on the standardized stepwise regression, the relative gene expression of JA was significantly affected by the DInc and the PCA. The heatmap analysis provided the same results (Figure 6).

In general, antagonism occurs between the JA and SA biosynthetic pathways. Thus, if the relative gene expression of SA is high, the relative gene expression of JA is low. These hormonal signaling pathways do not act independently but affect each other through a complex network of regulatory interactions [35]. Several researchers have reported synergistic interactions between the JA and SA biosynthetic pathways [36,37]. Arbuscular mycorrhizal symbioses increased the production of SA. The increase was not dependent on the genotype's responsiveness to mycorrhizae, but it depended on their resistance to *R. solanacearum*. The results of the standardized stepwise regression reinforced this result. The relative gene expression of SA was significantly affected by DInc., either the mycorrhizal inoculation only (M1R0) or both arbuscular mycorrhizal and *R. solanacearum* (M1R1) inoculation.

It is known that the presence of endophytic mycorrhiza can mediate plant resistance by changing endogenous hormones. Symbiosis with endophytic fungi is usually associated with inhibiting the SA pathway and increasing JA production [19,38]. The results showed a significant increase in the relative gene expression of the JA and SA pathways, indicating that mycorrhizal colonization of the tropical chili pepper genotypes increased plant defense. Similarly, the tropical chili pepper genotypes inoculated with mycorrhizae increased their relative gene expression of JA, and the increase was higher than that of SA (Figures 2 and 3). The expression of JA and SA were in the opposite way. When JA was expressed higher, the SA would be low, and vice versa. Mycorrhizal colonization induced gene expression, activated the JA pathway, and increased plant resistance to *R. solanacearum*. JA levels increased significantly, particularly in the genotypes inoculated with mycorrhizae and infected with *R. solanacearum* (M1R1). Figures 7 and 8 showed a comparison of plant height and stem diameter of chili pepper plants with and without mycorrhiza inoculation. It may represent an increase in plant resistance to *R. solanacearum*. The chili pepper plant inoculated with mycorrhiza and *R. solanacearum* (M1R1) showed a higher plant height and thicker stem diameter compared to the chili pepper plant inoculated with *R. solanacearum* but without mycorrhiza (M0R1). Mycorrhizal colonization protected the tropical chili pepper from *R. solanacearum* and enhanced the immune priming system by producing JA and SA in genotypes with high and low responsiveness or hostile responses to mycorrhizae. However, the increase in the relative gene expression of the JA and SA signaling pathways was affected by the resistance properties of the tropical chili pepper genotypes to *R. solanacearum*.

As biotrophic organisms, mycorrhiza can trigger plant defense responses during the early stages of colonization, similar to biotrophic pathogen infection [67,70]. Plant responses occur in defense of colonization. Modulating plant defense responses leads to tissue preconditioning to activate plant defenses systemically during a pathogen attack by activating JA signaling, which is referred to as priming. Plant responses are faster and more robust in dealing with a pathogen infection when primed so they are more efficient at increasing plant resistance and plant health [66,67,71,72]. This priming phenomenon was found in the mycorrhizae and *R. solanacearum* (M1R1) inoculated treatment, especially in the genotypes resistant to *R. solanacearum*. The relative gene expression of JA increased in the five genotypes of tropical chili peppers in the M1R1 treatment, regardless of the responsiveness of the chili pepper to the mycorrhizae (Figure 2).

Resistance to *R. solanacearum* is closely related to and controlled by JA and SA-mediated signaling [73]. Mycorrhizae potentially control plant diseases, particularly soil-borne pathogens [26–28]. The presence of mycorrhizae induces the reprogramming of plant metabolic pathways involved in defense, which may increase the ability of the plants to survive adverse conditions [5,19,23–25]. Local and systemic MIR increases the content of plant hormones, such as SA and JA [23,24,29–31].

Symbiosis with mycorrhizae increases the root surface area more than 100 times compared to non-symbiotic roots [74]. Colonization with arbuscular mycorrhizae changes the root morphology, and these changes make it difficult for soil-borne pathogens to enter the root system. The symbiosis directly causes an increase in mycorrhizal exudation from root branches, which changes the root microbiome and stimulates root exudates that inhibit pathogens. This also increases the acquisition of mineral nutrients and modulates plant hormonal balance. Arbuscular mycelia protect the roots from the damaging effects of various pathogenic hydrolytic enzymes that disrupt the epidermal cell walls. Arbuscular mycelia block the molecular crosstalk required for the host plant/pathogen interaction with the rhizosphere [23,24,75].

5. Conclusions

Among the eight accessions, three groups were resistant to *R. solanacearum*. Three genotypes were resistant, one genotype was moderately resistant, and four genotypes were susceptible to *R. solanacearum*. Based on the responsiveness to the arbuscular mycorrhizae, the ten genotypes were classified as four tropical chili pepper genotypes with high responsiveness and six with low responsiveness to the arbuscular mycorrhizae, three of which exhibited negative responsiveness. Inoculating the tropical chili peppers with arbuscular mycorrhizae induced resistance to *R. solanacearum* through a priming action by increasing the relative expression of the genes associated with JA and SA. However, the relationship was different between the JA and SA pathways. Mycorrhizae induced the relative expression of the gene associated with JA in the tropical chili peppers, which depended on the responsiveness to mycorrhizae and resistance to *R. solanacearum*. The relative expression of the gene associated with SA was depended on the resistance of the genotypes to *R. solanacearum*. These results suggest that mycorrhizal colonization modulates the signaling pathways mediated by JA and SA to activate plant defense genes against bacterial wilt caused by *R. solanacearum* and increase plant resistance.

Author Contributions: Conceptualization, E.A., T., T.A. (Triwidodo Arwiyanto) and J.W.; methodology, E.A., T., T.A. (Triwidodo Arwiyanto) and J.W.; software, E.A. and T.A. (Taufan Alam); validation, E.A., T., T.A. (Triwidodo Arwiyanto) and J.W.; formal analysis, E.A., T.A. (Taufan Alam) and I.P.A.; investigation, E.A.; resources, E.A.; data curation, E.A.; writing of the original draft, E.A.; writing, reviewing, and editing, E.A., T., T.A. (Triwidodo Arwiyanto) and J.W.; visualization, E.A., T.A. (Taufan Alam) and I.P.A.; supervision, T., T.A. (Triwidodo Arwiyanto) and J.W.; project administration, E.A. and T.A. (Taufan Alam); funding acquisition, E.A. and T. All authors have read and agreed to the published version of the manuscript.

Funding: This research was funded by the Universitas Gadjah Mada, grant number 3550/UN1.P.III/Dit-Lit/PT.01.05/2022.

Institutional Review Board Statement: Not applicable for studies not involving humans or animals.

Informed Consent Statement: Informed consent was obtained from all subjects involved in the study.

Data Availability Statement: The data presented in this study are available on request from the corresponding author.

Acknowledgments: Our profound gratitude is addressed to Woro Darmini, who assisted in culturing the bacterial suspension of *R. solanacearum*, and Yeni Fatmawati and Putri Eka Amdella, who assisted in the preparation and operation of the qRT-PCR equipment.

Conflicts of Interest: The authors declare no conflict of interest.

References

1. Bosland, P.W.; Votava, E.J. *Peppers: Vegetable and Spice Capsicums*, 2nd ed.; CABI Publishing: Willingford, UK; Volume 2, pp. 1–230. [CrossRef]
2. López, P.; Gorzalczany, S.; Acevedo, C.; Alonso, R.; Ferraro, G. Chemical Study and Anti-Inflammatory Activity of *Capsicum chacoense* and *C. baccatum*. *Rev. Bras. Farmacogn.* **2012**, *22*, 455–458. [CrossRef]
3. Pereira, J.A.P.; Vieira, I.J.C.; Freitas, M.S.M.; Prins, C.L.; Martins, M.A.; Rodrigues, R. Effects of Arbuscular Mycorrhizal Fungi on *Capsicum* Spp. *J. Agric. Sci.* **2016**, *154*, 828–849. [CrossRef]
4. Sanati, S.; Razavi, B.M.; Hosseinzadeh, H. A Review of the Effects of *Capsicum annuum* L. and Its Constituent, Capsaicin, in Metabolic Syndrome. *Iran. J. Basic Med. Sci.* **2018**, *21*, 439–448. [CrossRef] [PubMed]
5. Ahmed, W.; Yang, J.; Tan, Y.; Munir, S.; Liu, Q.; Zhang, J.; Ji, G.; Zhao, Z. *Ralstonia solanacearum*, a Deadly Pathogen: Revisiting the Bacterial Wilt Biocontrol Practices in Tobacco and Other Solanaceae. *Rhizosphere* **2022**, *21*, 100479. [CrossRef]
6. Aguk, J.A.; Karanja, N.; Schulte-Geldermann, E.; Bruns, C.; Kinyua, Z.; Parker, M. Control of Bacterial Wilt (*Ralstonia solanacearum*) in Potato (*Solanum tuberosum*) Using Rhizobacteria and Arbuscular Mycorrhiza Fungi. *Afr. J. Food Agric. Nutr. Dev.* **2018**, *18*, 13371–13387. [CrossRef]
7. Du, H.; Chen, B.; Zhang, X.; Zhang, F.; Miller, S.A.; Rajashekara, G.; Xu, X.; Geng, S. Evaluation of *Ralstonia solanacearum* Infection Dynamics in Resistant and Susceptible Pepper Lines Using Bioluminescence Imaging. *Plant Dis.* **2017**, *101*, 272–278. [CrossRef]
8. Kang, Y.J.; Ahn, Y.K.; Kim, K.T.; Jun, T.H. Resequencing of *Capsicum annuum* Parental Lines (YCM334 and Taean) for the Genetic Analysis of Bacterial Wilt Resistance. *BMC Plant Biol.* **2016**, *16*, 235. [CrossRef]
9. Mamphogoro, T.P.; Babalola, O.O.; Aiyegoro, O.A. Sustainable Management Strategies for Bacterial Wilt of Sweet Peppers (*Capsicum annuum*) and Other Solanaceous Crops. *J. Appl. Microbiol.* **2020**, *129*, 496–508. [CrossRef]
10. Thakur, P.P.; Mathew, D.; Nazeem, P.A.; Abida, P.S.; Indira, P.; Girija, D.; Shylaja, M.R.; Valsala, P.A. Identification of Allele-Specific AFLP Markers Linked with Bacterial Wilt [*Ralstonia solanacearum* (Smith) Yabuuchi et al.] Resistance in Hot Peppers (*Capsicum annuum* L.). *Physiol. Mol. Plant Pathol.* **2014**, *87*, 19–24. [CrossRef]
11. Mansfield, J.; Genin, S.; Magori, S.; Citovsky, V.; Sriariyanum, M.; Ronald, P.; Dow, M.; Verdier, V.; Beer, S.V.; Machado, M.A.; et al. Top 10 Plant Pathogenic Bacteria in Molecular Plant Pathology. *Mol. Plant Pathol.* **2012**, *13*, 614–629. [CrossRef]
12. Lebeau, A.; Daunay, M.C.; Frary, A.; Palloix, A.; Wang, J.F.; Dintinger, J.; Chiroleu, F.; Wicker, E.; Prior, P. Bacterial Wilt Resistance in Tomato, Pepper, and Eggplant: Genetic Resources Respond to Diverse Strains in the *Ralstonia solanacearum* Species Complex. *Phytopathology* **2011**, *101*, 154–165. [CrossRef]
13. Álvarez, B.; Biosca, E.G.; López, M.M. On the Life of *Ralstonia solanacearum*, a Destructive Bacterial Plant Pathogen. In *Technology and Education Topics in Applied Microbiology and Microbial Biotechnology*; Formatex Research Center: Badajoz, Spain, 2010; pp. 267–279.

14. Yadeta, K.A.; Thomma, B.P.H.J. The Xylem as a Battleground for Plant Hosts and Vascular Wilt Pathogens. *Front. Plant Sci.* **2013**, *4*, 97. [CrossRef]
15. Jiang, G.; Wei, Z.; Xu, J.; Chen, H.; Zhang, Y.; She, X.; Macho, A.P.; Ding, W.; Liao, B. Bacterial Wilt in China: History, Current Status, and Future Perspectives. *Front. Plant Sci.* **2017**, *8*, 1549. [CrossRef] [PubMed]
16. Agoncillo, E.S. Control Bacterial Wilt Disease Caused by *Ralstonia Solanacearum* in Pepper Using Arbuscular Mycorrhizal Fungi (Mykovam). *J. Nat. Sci. Res.* **2018**, *8*, 62–66.
17. Namisy, A.; Chen, J.R.; Prohens, J.; Metwally, E.; Elmahrouk, M.; Rakha, M. Screening Cultivated Eggplant and Wild Relatives for Resistance to Bacterial Wilt (*Ralstonia Solanacearum*). *Agriculture* **2019**, *9*, 157. [CrossRef]
18. Kim, B.-S.; French, E.; Caldwell, D.; Harrington, E.J.; Iyer-Pascuzzi, A.S. Bacterial Wilt Disease: Host Resistance and Pathogen Virulence Mechanisms. *Physiol. Mol. Plant Pathol.* **2016**, *95*, 37–43. [CrossRef]
19. Llorens, E.; Scalschi, L.; Sharon, O.; Vicedo, B.; Sharon, A.; García-Agustín, P. Jasmonic Acid Pathway Is Required in the Resistance Induced by *Acremonium sclerotigenum* in Tomato against *Pseudomonas syringae*. *Plant Sci.* **2022**, *318*, 111210. [CrossRef]
20. Yuan, S.; Li, M.; Fang, Z.; Liu, Y.; Shi, W.; Pan, B.; Wu, K.; Shi, J.; Shen, B.; Shen, Q. Biological Control of Tobacco Bacterial Wilt Using *Trichoderma harzianum* Amended Bioorganic Fertilizer and the Arbuscular Mycorrhizal Fungi *Glomus mosseae*. *Biol. Control* **2016**, *92*, 164–171. [CrossRef]
21. Latz, M.A.C.; Jensen, B.; Collinge, D.B.; Jørgensen, H.J.L. Endophytic Fungi as Biocontrol Agents: Elucidating Mechanisms in Disease Suppression. *Plant Ecol. Divers.* **2018**, *11*, 555–567. [CrossRef]
22. Enebe, M.C.; Babalola, O.O. The Impact of Microbes in the Orchestration of Plants' Resistance to Biotic Stress: A Disease Management Approach. *Appl. Microbiol. Biotechnol.* **2019**, *103*, 9–25. [CrossRef]
23. Dowarah, B.; Gill, S.S.; Agarwala, N. Arbuscular Mycorrhizal Fungi in Conferring Tolerance to Biotic Stresses in Plants. *J. Plant Growth Regul.* **2021**, *41*, 1429–1444. [CrossRef]
24. Sanmartín, N.; Pastor, V.; Pastor-Fernández, J.; Flors, V.; Pozo, M.J.; Sánchez-Bel, P. Role and Mechanisms of Callose Priming in Mycorrhiza-Induced Resistance. *J. Exp. Bot.* **2021**, *71*, 2769–2781. [CrossRef] [PubMed]
25. Singh, A.; Gupta, R.; Pandey, R. Rice Seed Priming with Picomolar Rutin Enhances Rhizospheric *Bacillus subtilis* CIM Colonization and Plant Growth. *PLoS ONE* **2016**, *11*, e01460132. [CrossRef] [PubMed]
26. Whipps, J.M. Prospects and Limitations for Mycorrhizas in Biocontrol of Root Pathogens. *Can. J. Bot.* **2004**, *82*, 1198–1227. [CrossRef]
27. Jacott, C.N.; Murray, J.D.; Ridout, C.J. Trade-Offs in Arbuscular Mycorrhizal Symbiosis: Disease Resistance, Growth Responses and Perspectives for Crop Breeding. *Agronomy* **2017**, *7*, 75. [CrossRef]
28. Bonfante, P.; Genre, A. Mechanisms Underlying Beneficial Plant—Fungus Interactions in Mycorrhizal Symbiosis. *Nat. Commun.* **2010**, *1*, 48. [CrossRef] [PubMed]
29. Garcia-Garrido, J.M. Regulation of the Plant Defence Response in Arbuscular Mycorrhizal Symbiosis. *J. Exp. Bot.* **2002**, *53*, 1377–1386. [CrossRef]
30. Hause, B.; Mrosk, C.; Isayenkov, S.; Strack, D. Jasmonates in Arbuscular Mycorrhizal Interactions. *Phytochemistry* **2007**, *68*, 101–110. [CrossRef]
31. Hohmann, P.; Messmer, M.M. Breeding for Mycorrhizal Symbiosis: Focus on Disease Resistance. *Euphytica* **2017**, *213*, 113. [CrossRef]
32. Llorens, E.; García-Agustín, P.; Lapeña, L. Advances in Induced Resistance by Natural Compounds: Towards New Options for Woody Crop Protection. *Sci. Agric.* **2017**, *74*, 90–100. [CrossRef]
33. Koo, A.J.K.; Howe, G.A. The Wound Hormone Jasmonate. *Phytochemistry* **2009**, *70*, 1571–1580. [CrossRef] [PubMed]
34. De Geyter, N.; Gholami, A.; Goormachtig, S.; Goossens, A. Transcriptional Machineries in Jasmonate-Elicited Plant Secondary Metabolism. *Trends Plant Sci.* **2012**, *17*, 349–359. [CrossRef] [PubMed]
35. Pieterse, C.M.J.; Van Der Does, D.; Zamioudis, C.; Leon-Reyes, A.; Van Wees, S.C.M. Hormonal Modulation of Plant Immunity. *Annu. Rev. Cell Dev. Biol.* **2012**, *28*, 489–521. [CrossRef] [PubMed]
36. Halim, V.A.; Altmann, S.; Ellinger, D.; Eschen-Lippold, L.; Miersch, O.; Scheel, D.; Rosahl, S. PAMP-Induced Defense Responses in Potato Require Both Salicylic Acid and Jasmonic Acid. *Plant J.* **2009**, *57*, 230–242. [CrossRef]
37. Scalschi, L.; Vicedo, B.; Camañes, G.; Fernandez-Crespo, E.; Lapeña, L.; González-Bosch, C.; García-Agustín, P. Hexanoic Acid Is a Resistance Inducer That Protects Tomato Plants against Pseudomonas Syringae by Priming the Jasmonic Acid and Salicylic Acid Pathways. *Mol. Plant Pathol.* **2013**, *14*, 342–355. [CrossRef]
38. Navarro-Meléndez, A.L.; Heil, M. Symptomless Endophytic Fungi Suppress Endogenous Levels of Salicylic Acid and Interact with the Jasmonate-Dependent Indirect Defense Traits of Their Host, Lima Bean (*Phaseolus lunatus*). *J. Chem. Ecol.* **2014**, *40*, 816–825. [CrossRef]
39. Hetrick, B.A.D.; Wilson, G.W.T.; Cox, T.S. Mycorrhizal Dependence of Modern Wheat Cultivars and Ancestors: A Synthesis. *Can. J. Bot.* **1992**, *71*, 512–518. [CrossRef]
40. Phillips, J.M.; Hayman, D.S. Improved Procedures for Clearing Roots and Staining Parasitic and Vesicular-Arbuscular Mycorrhizal Fungi for Rapid Assessment of Infection. *Trans. Br. Mycol. Soc.* **1970**, *55*, 158–161. [CrossRef]
41. Schindelin, J.; Arganda-Carreras, I.; Frise, E.; Kaynig, V.; Longair, M.; Pietzsch, T.; Preibisch, S.; Rueden, C.; Saalfeld, S.; Schmid, B.; et al. Fiji: An Open-Source Platform for Biological-Image Analysis. *Nat. Methods* **2012**, *9*, 676–682. [CrossRef]

42. Lafortune, D.; Béramis, M.; Daubèze, A.M.; Boissot, N.; Palloix, A. Partial Resistance of Pepper to Bacterial Wilt Is Oligogenic and Stable under Tropical Conditions. *Plant Dis.* **2005**, *89*, 501–506. [CrossRef]
43. Lillemo, M.; Joshi, A.K.; Prasad, R.; Chand, R.; Singh, R.P. QTL for Spot Blotch Resistance in Bread Wheat Line Saar Co-Locate to the Biotrophic Disease Resistance Loci Lr34 and Lr46. *Theor. Appl. Genet.* **2013**, *126*, 711–719. [CrossRef] [PubMed]
44. Geneaid Biotech Ltd. Home Page. Available online: https://www.geneaid.com/ (accessed on 16 March 2022).
45. Livak, K.J.; Schmittgen, T.D. Analysis of Relative Gene Expression Data Using Real-Time Quantitative PCR and the $2^{-\Delta\Delta CT}$ Method. *Methods* **2001**, *25*, 402–408. [CrossRef] [PubMed]
46. He, Y.; Fukushige, H.; Hildebrand, D.F.; Gan, S. Evidence Supporting a Role of Jasmonic Acid in Arabidopsis Leaf Senescence. *Plant Physiol.* **2002**, *128*, 876–884. [CrossRef] [PubMed]
47. Fu, X.; Feng, Y.Q.; Zhang, X.W.; Zhang, Y.Y.; Bi, H.G.; Ai, X.Z. Salicylic Acid Is Involved in Rootstock–Scion Communication in Improving the Chilling Tolerance of Grafted Cucumber. *Front. Plant Sci.* **2021**, *12*, 693344. [CrossRef]
48. Team R Development Core. *R: A Language and Environment for Statistical Computing*; R Foundation for Statistical Computing: Vienna, Austria, 2018; Available online: https://www.R-project.org (accessed on 5 August 2021).
49. Ghozali, I. *Structural Equation Modeling; Alternative Method with Partial Least Square (PLS)*, 2nd ed.; Badan Penerbit Universitas Diponegoro: Semarang, Indonesia, 2008.
50. Fernandes, A. *Structural Equation Modeling Approach PLS and SEM; Application of Smart PLS and AMOS Software*; Laboratory of Statistics, Faculty of Mathematics and Natural Sciences, Brawijaya University: Malang, Indonesia, 2008.
51. Krall, J.; Uthoff, V.; Harley, J. A Step-up Procedure for Selecting Variables Associated with Survival. *Biometrics* **1975**, *31*, 49–57. [CrossRef]
52. SAS Institute. *SAS System for Windows 9.4*; SAS Institute, Inc.: Cary, NC, USA, 2013.
53. Raivo, K. Package "Pheatmap": Prety Heatmap. Available online: https://cran.r-project.org/web/packages/pheatmap/index.html (accessed on 6 September 2021).
54. Salloum, M.S.; Guzzo, M.C.; Velazquez, M.S.; Sagadin, M.B.; Luna, C.M. Variability in Colonization of Arbuscular Mycorrhizal Fungi and Its Effect on Mycorrhizal Dependency of Improved and Unimproved Soybean Cultivars. *Can. J. Microbiol.* **2016**, *62*, 1034–1040. [CrossRef]
55. Sawers, R.J.H.; Gebreselassie, M.N.; Janos, D.P.; Paszkowski, U. Characterizing Variation in Mycorrhiza Effect among Diverse Plant Varieties. *Theor. Appl. Genet.* **2010**, *120*, 1029–1039. [CrossRef]
56. Mishra, V.; Ellouze, W.; Howard, R.J. Utility of Arbuscular Mycorrhizal Fungi for Improved Production and Disease Mitigation in Organic and Hydroponic Greenhouse Crops. *J. Hortic.* **2018**, *5*, 1000237. [CrossRef]
57. Tawaraya, K. Arbuscular Mycorrhizal Dependency of Different Plant Species and Cultivars. *Soil Sci. Plant Nutr.* **2003**, *49*, 655–668. [CrossRef]
58. Janos, D.P. Plant Responsiveness to Mycorrhizas Differs from Dependence upon Mycorrhizas. *Mycorrhiza* **2007**, *17*, 75–91. [CrossRef]
59. Matsunaga, H.; Saito, T.; Saito, A. Evaluation of Resistance to Bacterial Wilt and Phytophthora Blight in Capsicum Genetic Resources Collected in Myanmar. *J. Jpn. Soc. Hortic. Sci.* **2011**, *80*, 426–433. [CrossRef]
60. Mimura, Y.; Kageyama, T.; Minamiyama, Y.; Hirai, M. QTL Analysis for Resistance to Ralstonia Solanacearum in Capsicum Accession "LS2341". *J. Jpn. Soc. Hortic. Sci.* **2009**, *78*, 307–313. [CrossRef]
61. Aslam, M.N.; Mukhtar, T.; Hussain, M.A.; Raheel, M. Assessment of Resistance to Bacterial Wilt Incited by *Ralstonia solanacearum* in Tomato Germplasm. *J. Plant Dis. Prot.* **2017**, *124*, 585–590. [CrossRef]
62. Matsunaga, H.; Monma, S. Sources of Resistance to Bacterial Wilt in Capsicum. *J. Jpn. Soc. Hort. Sci.* **1999**, *68*, 753–761. [CrossRef]
63. Pawaskar, J.; Kadam, J.; Navathe, S.; Kadam, J. Response of Chilli Varieties and Genotypes to Bacterial Wilt Caused by *Ralstonia solanacearum* and Its Management. *Indian J. Sci.* **2014**, *11*, 66–72.
64. Mimura, Y.; Yoshikawa, M.; Hirai, M. Pepper Accession LS2341 Is Highly Resistant to *Ralstonia solanacearum* Strains from Japan. *HortScience* **2009**, *44*, 2038–2040. [CrossRef]
65. Tripodi, P.; Kumar, S. The Capsicum Crop: An Introduction. In *Compendium of Plant Geomes*; Springer Nature Switzerland AG: Basel, Switzerland, 2019; pp. 1–8. [CrossRef]
66. Cameron, D.D.; Neal, A.L.; van Wees, S.C.M.; Ton, J. Mycorrhiza-Induced Resistance: More than the Sum of Its Parts? *Trends Plant Sci.* **2013**, *18*, 539–545. [CrossRef]
67. Jung, S.C.; Martinez-Medina, A.; Lopez-Raez, J.A.; Pozo, M.J. Mycorrhiza-Induced Resistance and Priming of Plant Defenses. *J. Chem. Ecol.* **2012**, *38*, 651–664. [CrossRef]
68. Liu, H.; Wu, M.; Liu, J.; Qu, Y.; Gao, Y.; Ren, A. Tripartite Interactions Between Endophytic Fungi, Arbuscular Mycorrhizal Fungi, and Leymus Chinensis. *Microb. Ecol.* **2020**, *79*, 98–109. [CrossRef]
69. Zhang, L.; Zhou, J.; George, T.S.; Limpens, E.; Feng, G. Arbuscular Mycorrhizal Fungi Conducting the Hyphosphere Bacterial Orchestra. *Trends Plant Sci.* **2021**, *27*, 402–411. [CrossRef]
70. Paszkowski, U. Mutualism and Parasitism: The Yin and Yang of Plant Symbioses. *Curr. Opin. Plant Biol.* **2006**, *9*, 364–370. [CrossRef]
71. Van Wees, S.C.; Van der Ent, S.; Pieterse, C.M. Plant Immune Responses Triggered by Beneficial Microbes. *Curr. Opin. Plant Biol.* **2008**, *11*, 443–448. [CrossRef] [PubMed]

72. Pozo, M.J.; Azcón-Aguilar, C. Unraveling Mycorrhiza-Induced Resistance. *Curr. Opin. Plant Biol.* **2007**, *10*, 393–398. [CrossRef] [PubMed]
73. Yang, S.; Shi, Y.; Zou, L.; Huang, J.; Shen, L.; Wang, Y.; Guan, D.; He, S. Pepper CaMLO6 Negatively Regulates *Ralstonia solanacearum* Resistance and Positively Regulates High Temperature and High Humidity Responses. *Plant Cell Physiol.* **2020**, *61*, 1223–1238. [CrossRef] [PubMed]
74. Smith, S.E.; Read, D.J. *Mycorrhizal Symbiosis*, 3rd ed.; Academic Press: London, UK, 2008; Available online: https://www.scirp.org/(S(351jmbntvnsjt1aadkozje))/reference/referencespapers.aspx?referenceid=1398632 (accessed on 5 August 2021).
75. Kaur, S.; Suseela, V. Unraveling Arbuscular Mycorrhiza-Induced Changes in Plant Primary and Secondary Metabolome. *Metabolites* **2020**, *10*, 335. [CrossRef]

horticulturae

Communication

Touch-Induced Transcriptional Changes in Flower Buds of a Non-Model Horticultural Plant *Dianthus hybrida*

Ryo Nishijima [1], Alvin Sanjaya [1], Harue Shinoyama [2] and Yusuke Kazama [1,*]

[1] Department of Bioscience and Biotechnology, Fukui Prefectural University, 4-1-1, Kenjojima, Eiheiji 910-1195, Japan

[2] Department of Sustainable Agri-Culture, Fukui Prefectural University, 88-1, Futaomote, Awara 910-4103, Japan

* Correspondence: ykaze@fpu.ac.jp

Abstract: Touch stimulus responses are common in plants. Some flowering plants sense the arrival of their pollinators and secrete nectar or release pollen sacs, facilitating successful pollination. Molecular mechanisms for mechanical stimulus responses in plants are well characterized in *Arabidopsis* leaves, but not in non-model plants or other organs such as flowers. Here, we performed RNA-seq analysis of touched flower buds of *Dianthus hybrida*, a major ornamental plant. Upon touch treatment, 931 and 132 genes were upregulated and downregulated, respectively. GO enrichment analysis revealed that genes encoding serine/threonine protein kinases were significantly abundant among the upregulated genes, which is consistent with previous studies that demonstrated the pivotal role of protein phosphorylation in the touch stimulus response of *Arabidopsis* leaves. In comparison with the gene expression profile of touched *Arabidopsis* leaves, the same families but different homologs of the representative touch-induced genes encoding protein kinases were upregulated, showing that phosphorelay signaling was the common mechanism for touch stimulus response in flowers and leaves, but the players of the phosphorelay signaling were different. These results will contribute to further studies on the mechanical stimulus responses of ornamental flowers and the utilization of this mechanism for breeding programs.

Keywords: tactile stimulus; ornamental plant; RNA-seq

1. Introduction

Plants continuously encounter and can sense environmental mechanical stimuli such as wind, rainfall, sounds, and touch to control their growth and development [1,2]. Mechanical stimuli contribute to the successful pollination of some flowers. For example, in the legume *Desmodium setigerum*, flower color and shape are changed by pollination and bee visits [3]. In the case of *Oenothera drummondii*, flowers produce nectar that is sweeter than usual after being treated with the playback sound of a flying bee [4]. The male flower of the dioecious *Catasetum* releases pollen sacs when pollinators touch the center of the flower, helping to effectively pollinate this species [5]. Although flowers are thought to sense mechanical stimuli, their response to them is less documented.

Molecular mechanisms of touch stimulus responses in plants have been studied exclusively in *Arabidopsis thaliana*. Many signaling molecules, including potential secondary messengers and plant hormones, have been implicated in touch-induced responses. First, a very rapid change in cytosolic free Ca^{2+} concentration occurs after the stimulation of plant cells [6–12]. An increase in Ca^{2+} concentrations is sensed by Ca^{2+}-binding proteins that regulate downstream molecular processes [8–12] such as the phosphorylation of proteins, resulting in the modulation of protein conformation [13–15]. In *Arabidopsis*, 24 touch-responsive phosphopeptides, including kinases, phosphatases, cytoskeleton proteins, membrane proteins, and ion transporters, were identified after initiation of touch treatments [16]. A non-phosphorylated isoform of touch-regulated phosphoprotein 1 (TREPH1) exhibited insensitive phenotypes to touch treatments and loss of the induction of touch-induced gene expression [16].

Touch-induced gene expression occurred 10 to 40 min after the initiation of touch treatments, while the increase in Ca^{2+} concentrations and Ca^{2+}-independent protein phosphorylation occurred shortly after treatment. The representative touch-induced genes, *TCHs*, were originally isolated as responsible genes after stimuli, such as touch, wind, rain, wounding, and darkness [1]. *TCH1* encodes an *Arabidopsis* calmodulin, CAM2, whereas *TCH2* and *TCH3* encode calmodulin-like (CML) proteins, CML24 and CML12, respectively [1,17–19], which is consistent with the above description that Ca^{2+}-dependent pathways are involved in touch stimulation. Touch-induced genes were also identified throughout the genome, and a representative gene, *CML39* was isolated [20]. *TCH4* encodes xyloglucan endotransglucosylase/hydrolase [21].

The phytohormone jasmonate (JA) is also an important molecule involved in touch-induced responses in plants. An *Arabidopsis* mutant defective in allene oxide synthase (*aos*) and JA did not show touch-induced morphological changes [22]. In contrast, the expression of touch-induced genes was detected in the mutant, indicating that this was not sufficient to show touch-induced morphological changes, and that the expression of these genes was independent of the JA signaling pathway or upstream of the pathway [22].

Although these characteristics have been well documented in *Arabidopsis* leaves, touch-induced gene expression in flowers remains mostly unknown. Here, we investigated the gene expression profiles after exposure to touch stimulus of the flowers of an ornamental plant, *Dianthus hybrida*, which is an interspecific hybrid between *D. chinensis* and *D. barbatus*. The genus *Dianthus*, a member of the family Caryophyllaceae [23], is one of the major ornamental flowers of commercial importance, with production similar to that of *Chrysanthemum* and *Rosa*. The *Dianthus* is commonly known as "carnation," which refers to *Dianthus caryophyllus* and several intra/interspecific hybrids. *Dianthus* flowers vary widely in colors and shapes [24–27]. Molecular understanding of physiological mechanism underlying environmental stimulus effects on flower morphology will facilitate to develop new varieties. The genome project of *D. caryophyllus* is almost complete [28], allowing us to perform transcriptome analyses. In this report, we aimed to list the differentially expressed genes after touch stimulation in *D. hybrida* flowers and indicate directions for future research that will serve as a basis for further development and utilization of this response in breeding programs.

2. Materials and Methods

2.1. Plant Materials and RNA Extraction

D. hybrida cultivar "Telstar Scarlet" plants were grown in plastic pots (10.5 cm diameter and 9.5 cm height) in a wind- and vibration-free PVC box placed in a growth chamber under the following conditions; 14 h light/10 h dark, temperature of 28 °C and relative humidity of 70% (Figure 1A). Photosynthetic photon flux density (PPFD) was 315 μmol m^{-2} s^{-1}. Three untouched flower buds were sampled as controls, and another three buds were touched by hand for 40 s with hands and sampled 40 min later, following a previous study on *Arabidopsis* leaves [20]. All biological replicates were sampled from different plants. Total RNA was extracted using the RNeasy Plant Mini Kit (QIAGEN, Germantown, MD, USA) following the manufacturer's instructions.

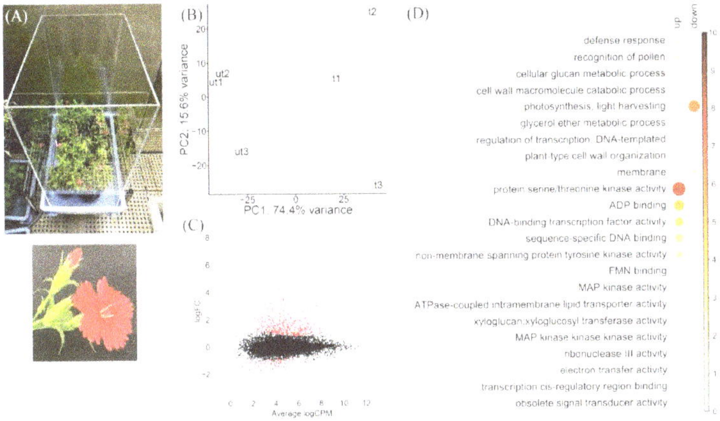

Figure 1. RNA-seq summary of touched *D. hybrida* flower buds. (**A**) *D. hybrida* plants grown in a wind- and vibration-proofed box. The bottom panel shows a flower and a bud. (**B**) PCA plot of three replicates each of touched (t) and untouched (ut) samples using uniquely mapped reads. (**C**) MA plot of expressed genes (TPM > 2.5 at least in all three replicates of either touched or untouched sample). Red dots indicate DEGs (FDR < 0.001). (**D**) GO enrichment analysis of DEGs.

2.2. RNA-seq Analysis

Total RNA sequencing was conducted by Rhelixa (Tokyo, Japan). Strand-specific RNA-seq libraries were prepared using the NEBNext Ultra II Directional mRNA-seq Kit for Illumina (New England Biolabs, Hitchin, Hertfordshire, UK). The libraries were bulked with others and sequenced using the NovaSeq6000 (Illumina, San Diego, CA, USA) with paired-end 150-bp reads.

Due to unavailability of chromosome-level genome assembly of *D. hybrida*, the genome scaffold of the closely related species *Dianthus caryophyllus* (carnation), the highest-quality genome sequence within the genus, was used as a reference. Scaffold, gene annotation, and *Arabidopsis* homolog data for carnation were downloaded from the Carnation DB [28] (http://carnation.kazusa.or.jp, accessed on 7 December 2021). RNA-seq reads were mapped to carnation scaffolds using the STAR aligner ver. 2.7.3a [29] with options "–outFilterMultimapNmax 1 –quantMode GeneCounts". Differentially expressed genes (DEGs) were called with glmLRT of R package edgeR ver. 3.26.8 [30], and transcripts per kilobase million (TPM) values were computed after trimmed mean of M values (TMM) normalization. GO enrichment analysis was performed using the R package clusterProfiler ver 3.12.0 [31] and visualized using the R package corrplot ver 0.92 [32]. To compare the transcriptome profile of *D. hybrida* flower buds with that of *Arabidopsis* leaves, microarray data from a previous study [20] was used.

3. Results and Discussion

To explore the gene expression profiles in the flower buds of *D. hybrida* in response to touch stimulus, we performed transcriptome analysis of the flower buds 40 min after touch treatment. On average, 22.8 million reads were obtained for each sample, and 71.5% of the total reads were uniquely mapped to carnation scaffolds (Table S1). Transcripts from 15,258 genes were detected (TPM $\geq$ 2.5, in all three replicates in either touched or untouched conditions). Principal component analysis based on the uniquely mapped read count showed that touched and untouched samples were clearly separated by the PC1 value with a 74.4% contribution rate (Figure 1B), suggesting touch-induced transcriptional changes in the flower buds. Genes with the highest loading values along the PC2 axis were a *CEP1* homolog and three UDP-glycosyltransferase genes, which are involved in pollen development [33], indicating that the PC2 represented inflorescence developmental stage. Gene expression analysis showed that 961 and 132 genes were up- or downregulated by touch treatment, respectively (FDR < 0.001, edgeR; Figure 1C; Supplementary Excel File). GO enrichment analysis revealed that genes with GO "protein serine/threonine kinase activity" were significantly upregulated in response to touch stimuli (Figure 1D), which was consistent with previous studies that demonstrated the importance of protein phosphorylation in *Arabidopsis* touch-delayed bolting [16,34]. Other GO terms related to phosphorelay signaling, such as "non-membrane spanning protein tyrosine kinase activity", "MAP kinase activity", and "MAP kinase kinase kinase activity" were also enriched in touch-induced upregulated genes. In addition, GO "xyloglucan:xyloglucosyl transferase activity," a representative mechanosensitive response [21], was detected. Although related GO terms were not enriched significantly, calmodulin genes, such as a *CAM5* homolog, and JA biosynthetic (*LOX1, 4, OPR3*, and *AOC3* homologs) and signaling genes (*JAZ1* and *NINJA* homologs) were upregulated. On the other hand, genes with GO "photosynthesis" were significantly downregulated upon touch stimulus, which resembled responses to other biotic stresses including bacteria, insects, and wounding [35].

To explore whether touch stimulus responses are generally conserved between flowers and leaves, we compared the RNA-seq data of *D. hybrida* flower buds with the microarray data of *Arabidopsis* leaves from a previous study [20]. All DEGs commonly detected in both experiments showed the same expression changes (Figure 2A,B), except for one clock-related gene, *FKF1*, which appeared to reflect the difference in sampling time. *TCH4*, which encodes xyloglucan endotransglucosylase [21], and its two *Dianthus* homologs were commonly upregulated. For calmodulin, *CML49*, but not *CAM5*, was upregulated in touched *Arabidopsis* leaves, while their homologs, *Dca25393* and *Dca20630*, respectively, showed the opposite expression patterns in the flower buds of *D. hybrida*, which implies that different organs utilize different calmodulin family members as touch signaling components. GO enrichment analysis revealed that genes for "protein serine/threonine kinase activity" were enriched in commonly detected and flower bud-specific DEGs, but not in leaf-specific ones (Figure 2C). In fact, 32 out of 44 genes encoding serine/threonine kinases were detected as DEGs only in flower buds, while no kinase gene was differentially expressed specifically in leaves. These kinases might be key players in a flower-specific touch stimulus response. Although it must be considered that these DEGs specific to one condition might reflect the difference between the two species and/or growth condition, these results suggest that an organ-specific molecular mechanism for touch stimulus response might exist, which could be the basis of organ function, as in the case of flowers listening to bees [4].

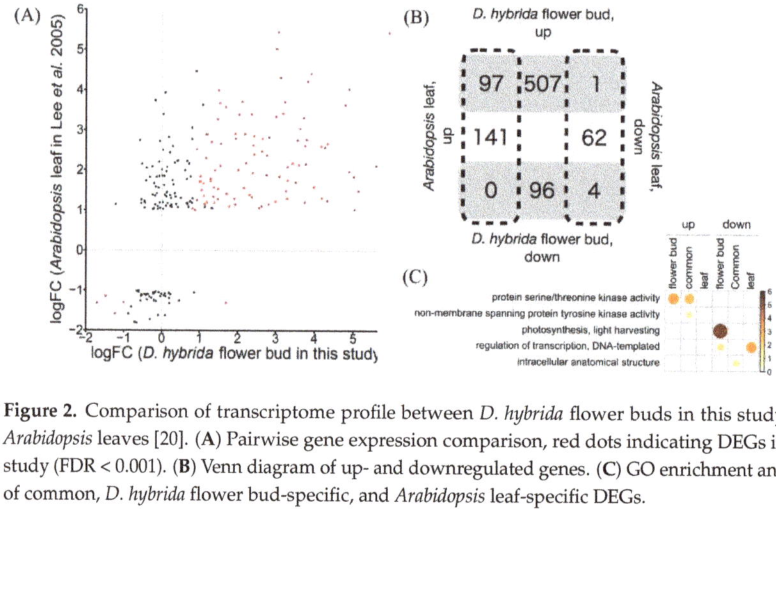

Figure 2. Comparison of transcriptome profile between *D. hybrida* flower buds in this study and *Arabidopsis* leaves [20]. (**A**) Pairwise gene expression comparison, red dots indicating DEGs in this study (FDR < 0.001). (**B**) Venn diagram of up- and downregulated genes. (**C**) GO enrichment analysis of common, *D. hybrida* flower bud-specific, and *Arabidopsis* leaf-specific DEGs.

4. Conclusions

Touch stimuli induced upregulation of genes encoding protein kinase, xyloglucan endotransglucosylase, calmodulin, and JA biosynthetic and signaling components and downregulation of photosynthetic genes in *D. hybrida* flower buds. This response was consistent with that of *Arabidopsis* leaves, in which protein phosphorylation events occur immediately after mechanical stimulation, followed by the increase in Ca^{2+} and JA concentrations [16,20,22], implying that molecular mechanisms for the touch stimulus response are generally conserved among plant organs and among plant species. In contrast, a large proportion of protein kinase genes was induced only in flower buds, suggesting an organ-specific response to tactile stimulation. Our results will contribute to further studies on physiological mechanisms underlying touch stimulus response of ornamental flowers and utilization of this response in future breeding programs.

Supplementary Materials: The following are available online at https://www.mdpi.com/article/10.3390/horticulturae8100918/s1, Table S1: mapping rate of RNA-seq reads to carnation scaffold sequences, File S1: statistical summary of differentially expressed genes.

Author Contributions: H.S. and Y.K. designed the research; A.S. and Y.K. performed the experiments; R.N. analyzed the data; R.N. and Y.K. wrote the manuscript with contributions from all authors. All authors have read and agreed to the published version of the manuscript.

Funding: This research was funded by the Grant for Strategic Issue Research Promotion 03-04R020207 from Fukui Prefectural University to H.S. and Y.K., and JSPS KAKENHI grant number 21K20585 to R.N. and 22H05071 and 21KK0128 to Y.K.

Institutional Review Board Statement: Not applicable.

Informed Consent Statement: Not applicable.

Data Availability Statement: The RNA-seq data reported in this study have been deposited in the DDBJ database (accession No. DRA014742).

Conflicts of Interest: The authors declare no conflict of interest.

References

1. Braam, J.; Davis, R.W. Rain-, wind-, and touch-induced expression of calmodulin and calmodulin-related genes in *Arabidopsis*. *Cell* **1990**, *60*, 357–364. [CrossRef]
2. Braam, J. In touch: Plant responses to mechanical stimuli. *New Phytol.* **2005**, *165*, 373–389. [CrossRef] [PubMed]
3. Willmer, P.; Stanley, D.A.; Steijven, K.; Matthews, I.M.; Nuttman, C.V. Bidirectional flower color and shape changes allow a second opportunity for pollination. *Curr. Biol.* **2009**, *19*, 919–923. [CrossRef]
4. Veits, M.; Khait, I.; Obolski, U.; Zinger, E.; Boonman, A.; Goldshtein, A.; Saban, K.; Seltzer, R.; Ben-Dor, U.; Estlein, P.; et al. Flowers respond to pollinator sound within minutes by increasing nectar sugar concentration. *Ecol. Lett.* **2019**, *22*, 1483–1492. [CrossRef] [PubMed]
5. Nicholson, C.C.; Bales, J.W.; Palmer-Fortune, J.E.; Nicholson, R.G. Darwin's bee-trap: The kinetics of *Catasetum*, a new world orchid. *Plant Signal. Behav.* **2008**, *3*, 19–23. [CrossRef] [PubMed]
6. Knight, M.R.; Campbell, A.K.; Smith, S.M.; Trewavas, A.J. Transgenic plant aequorin reports the effects of touch and cold-shock and elicitors on cytoplasmic calcium. *Nature* **1991**, *352*, 524–526. [CrossRef]
7. Zhu, X.; Feng, Y.; Liang, G.; Liu, N.; Zhu, J.K. Aequorin-based luminescence imaging reveals stimulus- and tissue-specific Ca^{2+} dynamics in Arabidopsis Plants. *Mol. Plant* **2013**, *6*, 444–455. [CrossRef]
8. Harper, J.F.; Breton, G.; Harmon, A. Decoding Ca^{2+} Signals through plant protein kinases. *Annu. Rev. Plant Biol.* **2004**, *55*, 263–288. [CrossRef]
9. Kim, M.C.; Chung, W.S.; Yun, D.J.; Cho, M.J. Calcium and calmodulin-mediated regulation of gene expression in plants. *Mol. Plant* **2009**, *2*, 13–21. [CrossRef]
10. Luan, S. The CBL-CIPK network in plant calcium signaling. *Trends Plant Sci.* **2009**, *14*, 37–42. [CrossRef]
11. DeFalco, T.A.; Bender, K.W.; Snedden, W.A. Breaking the code: Ca^{2+} sensors in plant signalling. *Biochem. J.* **2009**, *425*, 27–40. [CrossRef] [PubMed]
12. Kudla, J.; Batistič, O.; Hashimoto, K. Calcium signals: The lead currency of plant information processing. *Plant Cell* **2010**, *22*, 541–563. [CrossRef] [PubMed]
13. Hunter, T. Protein kinases and phosphatases: The yin and yang of protein phosphorylation and signaling. *Cell* **1995**, *80*, 225–236. [CrossRef]

14. Soderling, T.R. The Ca^{2+}–calmodulin-dependent protein kinase cascade. *Trends Biochem. Sci.* **1999**, *24*, 232–236. [CrossRef]
15. Clapham, D.E. Calcium signaling. *Cell* **2007**, *131*, 1047–1058. [CrossRef]
16. Wang, K.; Yang, Z.; Qing, D.; Ren, F.; Liu, S.; Zheng, Q.; Liu, J.; Zhang, W.; Dai, C.; Wu, M.; et al. Quantitative and functional posttranslational modification proteomics reveals that TREPH1 plays a role in plant touch-delayed bolting. *Proc. Natl. Acad. Sci. USA* **2018**, *115*, E10265–E10274. [CrossRef]
17. Sistrunk, M.L.; Antosiewicz, D.M.; Purugganan, M.M.; Braam, J. Arabidopsis *TCH3* encodes a novel Ca^{2+} binding protein and shows environmentally induced and tissue-specific regulation. *Plant Cell* **1994**, *6*, 1553–1565. [CrossRef]
18. Khan, A.R.; Johnson, K.A.; Braam, J.; James, M.N.G. Comparative modeling of the three-dimensional structure of the calmodulin-related TCH2 protein from *Arabidopsis*. *Proteins* **1997**, *27*, 144–153. [CrossRef]
19. McCormack, E.; Braam, J. Calmodulins and related potential calcium sensors of *Arabidopsis*. *New Phytol.* **2003**, *159*, 585–598. [CrossRef]
20. Lee, D.; Polisensky, D.H.; Braam, J. Genome-wide identification of touch- and darkness-regulated *Arabidopsis* Genes: A focus on calmodulin-like and XTH genes. *New Phytol.* **2005**, *165*, 429–444. [CrossRef]
21. Rose, J.K.C.; Braam, J.; Fry, S.C.; Nishitani, K. The XTH family of enzymes involved in xyloglucan endotransglucosylation and endohydrolysis: Current perspectives and a new unifying nomenclature. *Plant Cell Physiol.* **2002**, *43*, 1421–1435. [CrossRef] [PubMed]
22. Chehab, E.W.; Yao, C.; Henderson, Z.; Kim, S.; Braam, J. Arabidopsis touch-induced morphogenesis is jasmonate mediated and protects against pests. *Curr. Biol.* **2012**, *22*, 701–706. [CrossRef] [PubMed]
23. Harbaugh, D.T.; Nepokroeff, M.; Rabeler, R.K.; McNeill, J.; Zimmer, E.A.; Wagner, W.L. A new lineage-based tribal classification of the family Caryophyllaceae. *Int. J. Plant Sci.* **2010**, *171*, 185–198. [CrossRef]
24. Onozaki, T.; Mato, M.; Shibata, M.; Ikeda, H. Differences in flower color and pigment composition among white carnation (*Dianthus caryophyllus* L.) Cultivars. *Sci. Hortic.* **1999**, *82*, 103–111. [CrossRef]
25. Okamura, M.; Yasuno, N.; Ohtsuka, M.; Tanaka, A.; Shikazono, N.; Hase, Y. Wide variety of flower-color and -shape mutants regenerated from leaf cultures irradiated with ion beams. *Nucl. Instrum. Methods Phys. Res. B* **2003**, *206*, 574–578. [CrossRef]
26. Okamura, M.; Umemoto, N.; Onishi, N. Breeding glittering carnations by an efficient mutagenesis system. *Plant Biotechnol.* **2012**, *29*, 209–214. [CrossRef]
27. Okamura, M.; Nakayama, M.; Umemoto, N.; Cano, E.A.; Hase, Y.; Nishizaki, Y.; Sasaki, N.; Ozeki, Y. Crossbreeding of a metallic color carnation and diversification of the peculiar coloration by ion-beam irradiation. *Euphytica* **2013**, *191*, 45–56. [CrossRef]
28. Yagi, M.; Kosugi, S.; Hirakawa, H.; Ohmiya, A.; Tanase, K.; Harada, T.; Kishimoto, K.; Nakayama, M.; Ichimura, K.; Onozaki, T.; et al. Sequence analysis of the genome of carnation (*Dianthus caryophyllus* L.). *DNA Res.* **2014**, *21*, 231–241. [CrossRef]
29. Dobin, A.; Davis, C.A.; Schlesinger, F.; Drenkow, J.; Zaleski, C.; Jha, S.; Batut, P.; Chaisson, M.; Gingeras, T.R. STAR: Ultrafast universal RNA-seq aligner. *Bioinformatics* **2013**, *29*, 15–21. [CrossRef]
30. Robinson, M.D.; McCarthy, D.J.; Smyth, G.K. EdgeR: A bioconductor package for differential expression analysis of digital gene expression data. *Bioinformatics* **2010**, *26*, 139–140. [CrossRef]
31. Yu, G.; Wang, L.G.; Han, Y.; He, Q.Y. ClusterProfiler: An R package for comparing biological themes among gene clusters. *Omics* **2012**, *16*, 284–287. [CrossRef] [PubMed]
32. Wei, T.; Simko, V.R. Package "Corrplot": Visualization of a Correlation Matrix (Version 0.92). Available online: https://github.com/taiyun/corrplot (accessed on 1 September 2022).
33. Zhang, D.; Liu, D.; Lv, X.; Wang, Y.; Xun, Z.; Liu, Z.; Li, F.; Lu, H. The cysteine protease CEP1, a key executor involved in tapetal programmed cell death, regulates pollen development in *Arabidopsis*. *Plant Cell* **2014**, *26*, 2939–2961. [CrossRef] [PubMed]
34. Darwish, E.; Ghosh, R.; Ontiveros-Cisneros, A.; Tran, H.C.; Petersson, M.; de Milde, L.; Broda, M.; Goossens, A.; van Moerkercke, A.; Khan, K.; et al. Touch signaling and thigmomorphogenesis are regulated by complementary CAMTA3- and JA-dependent pathways. *Sci. Adv.* **2022**, *8*, eabm2091. [CrossRef]
35. Bilgin, D.D.; Zavala, J.A.; Zhu, J.; Clough, S.J.; Ort, D.R.; Delucia, E.H. Biotic stress globally downregulates photosynthesis genes. *Plant Cell Environ.* **2010**, *33*, 1597–1613. [CrossRef] [PubMed]

horticulturae

MDPI

Article

Transcriptome and Physiological Analyses of Resistant and Susceptible Pepper (*Capsicum annuum*) to *Verticillium dahliae* Inoculum

Xinmin Huang [1], Liming He [1], Huimin Tan [1], Jiayi Liu [1], Qiucheng Qiu [2], Qidi Sun [2], Lejun Ouyang [1], Hanbing Han [1,*] and Qinqin He [1,*]

1 College of Biology and Food Engineering, Guangdong University of Petrochemical Technology, Maoming 525000, China; hxmscau@163.com (X.H.); heliming7799@163.com (L.H.); tanhuimin1109@163.com (H.T.); liujiayi23333@163.com (J.L.); ouyanglejun@gdupt.edu.cn (L.O.)

2 Maoming Maoshu Seed Industry Technology Co., Ltd., Maoming 525000, China; mszyqqc@126.com (Q.Q.); mszysqd@126.com (Q.S.)

* Correspondence: hhb05@126.com (H.H.); xiaohe1507@126.com (Q.H.)

Abstract: Pepper (*Capsicum annuum*) is a globally important vegetable, and Verticillium wilt is an important disease affecting peppers and is caused by *Verticillium dahliae*, which can severely reduce yields. However, the molecular mechanisms underlying the responses of pepper to infection by *V. dahliae* are largely unknown. We performed physiological and transcriptome analysis using resistant and susceptible pepper cultivars inoculated with *V. dahliae*. Compared to the susceptible cultivar MS66, the resistant cultivar MS72 retained higher chlorophyll content and lower malondialdehyde content after inoculation. At 3 days after inoculation (DAI), compared with MS66, 534 differentially expressed genes (DEGs) were identified in MS72. At 5 DAI, 2392 DEGs were identified in MS72 compared with MS66. The DEGs in MS72 were mainly enriched in the cell wall and photosynthesis-related Gene Ontology terms, as well as in pathways such as cutin, suberin, wax biosynthesis, phenylpropanoid biosynthesis, and photosynthesis. Using weighted gene co-expression network analysis, 36 hub genes involved in the resistance response were identified, including the transcription factor *bHLH93* (Capana04g000815), defense-like protein 1 (MSTRG.5904), and miraculin-like (Capana10g002167). Our findings contribute to a more comprehensive understanding of the response mechanism of pepper to *V. dahliae* inoculation, providing new avenues for improving pepper resistance through breeding programs.

Keywords: pepper; Verticillium wilt; *Verticillium dahliae*; chlorophyll content; malondialdehyde concentration; differentially expressed genes; weighted gene co-expression network analysis; hub genes

1. Introduction

Pepper (*Capsicum annuum*) is an annual or perennial plant of the Solanaceae family, native to Mexico and South America, and widely planted in tropical and subtropical regions worldwide [1]. Pepper is an economically important crop that is closely associated with human life. It not only serves as an essential fresh vegetable and seasoning in human diets, but is also an important industrial raw material in pharmaceutical, agricultural, and cosmetic industries [2,3]. Pepper is rich in vitamins, minerals, and antioxidants and has a unique flavor that can add color and taste to various dishes while providing health benefits [2,3]. Capsanthin and capsorubin extracted from pepper have important applications in the food, chemical, and agricultural industries [4]. According to data from the Food and Agriculture Organization of the United Nations (FAO) in 2022, China's pepper cultivation area of 759,817 ha exceeded one-third of the global pepper cultivation area (2,020,816 ha). However, the production of peppers is often affected by various abiotic and biotic constraints [5,6]. In particular, Verticillium wilt, caused by the fungus *Verticillium dahliae*, poses a serious threat to global pepper production [7–10].

Verticillium wilt is a disease caused by soil-borne pathogens, mainly including *V. dahliae*, *V. alboatrum*, *V. nigrescens*, *V. nubilum*, and *V. tricorpus*. Verticillium wilt was first reported in Sweden in 1918 [11]. Since then, Verticillium wilt has been observed and reported worldwide [10,12–14]. In 2014, Verticillium wilt was first detected in Gansu, China [15]. Verticillium wilt can affect over 400 plant species globally [16,17]. The *V. dahliae* in soil can infect peppers through root wounds. When the fungus enters the xylem vessels of the root, it rapidly reproduces and spreads along the xylem vessels to the aboveground parts of the peppers. As the disease progresses, symptoms such as leaf yellowing, progressive necrosis, leaf shedding, and delayed development appear, ultimately decreasing fruit yield and quality [8,13].

The prevention and control of *V. dahliae* is extremely difficult, as this fungus develops a resting structure known as microsclerotia, which can survive in soil for more than 14 years [18]. Presently, the prevention of pepper verticillium wilt is primarily dependent on chemical control methods; however, the environmental persistence and associated adverse effects of pesticides on human health remain a problem [19,20]. There are also ways of using beneficial bacteria for biological control, such as *Funneliformis mosseae* and *Bacillus siamensis*, to prevent and treat pepper Verticillium wilt caused by *V. dahliae* [21–23]; however, these methods have certain limitations, including unstable effectiveness and high costs. Generally, breeding cultivars that are resistant to *V. dahliae* in the pepper is considered to be the most effective control method [24,25]. Remarkably, certain pepper germplasms that are resistant to *V. dahliae* have been identified. Gurung et al. reported that eight pepper germplasms, including Grif 9073, PI 281396, PI 281397, PI 438666, PI 439292, PI 439297, PI 555616, and PI 594125, exhibited high resistance to *V. dahliae* VdCa59 and *V. dahliae* VdCf45 [9]. P.I. 215699 (a mixture of *C. baccatum* var. *microcarpum* and *C. annuum*), P.I. 535616 (*C. annuum*), and P.I. 555614 (*C. annuum*) also exhibit high resistance to *V. dahliae* Kleb [25]. Although multiple germplasms have been screened for resistance to *V. dahliae*, the resistance genes of these germplasms have not been cloned, and the resistance mechanism has not been elucidated.

The response of plants to pathogens is a complex process involving a series of biological reactions and physiological processes. In a study on the use of arbuscular mycorrhizal fungi (AMF) to prevent and control pepper Verticillium wilt, AMF were found to enhance pepper resistance by increasing the levels of new subtypes of acidic chitinase and superoxide dismutase (SOD), and by enhancing the activities of peroxidase and phenylalanine ammonia-lyase in peppers [26]. Another study revealed that AMF and salicylic acid (SA) increased resistance to Verticillium wilt by enhancing the activity of antioxidant enzymes such as SOD, ascorbate peroxidase, and catalase in chili plants [27]. In addition, studies have revealed that an increase in early phenolic compounds may be the reason why COA H (a formulation containing SA, soluble ammonium salts, and seawater extract from *Ascophyllum nodorum*) enhances the resistance of pepper seedlings to Verticillium wilt [28]. However, the molecular changes underlying these physiological responses of resistance to Verticillium wilt are unclear, and knowledge about the resistance mechanism of peppers to *V. dahliae* is limited.

In recent years, significant progress has been made using transcriptome technology to reveal the mechanisms underlying plant disease resistance. Using transcriptomics to study the response mechanisms of two different wheat varieties resistant to Fusarium head blight, Zhao et al. found that glutathione metabolism, phenylpropanoid biosynthesis, plant hormone signal transduction, and plant–pathogen interactions in the Kyoto Encyclopedia of Genes and Genomes (KEGG) were related to disease resistance. Additionally, five pathogenesis-related proteins were found to be involved in resistance regulation [29]. After inoculation with *Coniella diprodella*, transcriptome analysis of the leaves of the two grape varieties by Li et al. showed that the defense mechanisms of resistant varieties involved plant and pathogen interactions, sulfur transfer systems, and biosynthesis of sesquiterpenes and triterpenes, suberin, wax, and monoterpenes, and biosynthesis pathways of flavonoids and flavonols [30]. Transcriptome analysis showed that the phenylpropanoid biosynthesis pathway was activated in cotton after inoculation with *V. dahliae*. Upon combining the results of this analysis with metabolomics data, five phenylpropane metabolites, caffeic acid, coniferyl alcohol, coniferin, scopoletin, and scopoline, were identified. In vitro experiments showed that they can significantly inhibit the growth of *V. dahliae*, indicating that they may play an important role in cotton's resistance to *V. dahliae* [31]. Recent studies have shown that the endophytic fungus *Gibellulopsis nigrescens* CEF08111 can significantly enhance pathways involved in plant–pathogen interactions, mitogen-activated protein kinase (MAPK) signaling, and plant hormone signal transduction, thereby improving cotton resistance to Verticillium wilt after 12 h of inoculation. The genes encoding calcium-dependent protein kinase (CDPK), flagellin sensing 2 (FLS2), resistance to *Pseudomonas syringae* pv. *maculicola* 1 (RPM1), and mitosis protein 2 (MYC2) may be involved in regulating resistance to cotton Verticillium wilt [32].

To understand the molecular-level changes of different resistant materials in peppers after infection with *V. dahliae*, we conducted physiological and transcriptome analysis at three different time points after inoculation with *V. dahliae* in both resistant and susceptible cultivars. We identified the DEGs of these two cultivars and screened candidate genes for pepper resistance to *V. dahliae*. These results provide a foundation for understanding the response mechanism of peppers to *V. dahliae*, and offer valuable information for breeding disease-resistant pepper cultivars.

2. Materials and Methods

2.1. Plant Materials and Treatment

The pepper cultivars MS66 and MS72 used in this experiment were selected and provided by Maoming Maoshu Seed Industry Technology Co., Ltd. (Maoming, China). Our previous research has shown that MS66 is susceptible to *V. dahliae*, resulting in yellowing and wilting of leaves after inoculation, whereas MS72 is resistant to *V. dahliae*, with minimal leaf yellowing observed after inoculation. Peppers were sown on 5 October 2023, and seedlings were planted in a polyethylene film greenhouse (20–28 °C; natural sunlight; 11–12 h of sunlight per day) at Guangdong University of Petrochemical Technology. On 17 January 2024, leaves with similar shapes and sizes, no gaps, healthy physiological status, and no pests or diseases were selected for inoculation with *V. dahliae*. *V. dahliae* was grown on potato dextrose agar (PDA, Huankai, Guangzhou, China) dishes at 24 °C for 5 days, in the dark. A single colony was selected and incubated in sucrose sodium nitrate liquid medium (Huankai) at 24 °C for 5 days in the dark. Then, a suspension of 10^7 conidia/mL was prepared from the culture using distilled sterile water [33]. The petioles of detached leaves were placed in a spore suspension. The detached leaves were stored at a temperature of 24 ± 2 °C, with a day–night light cycle of 16/8 h [34,35]. Samples of 30 leaves with three replicates were collected before inoculation (day 0) and 3 and 5 days after inoculation (DAI), and labelled MS66-0, MS66-3, and MS66-5; and MS72-0, MS72-3, and MS72-5 for MS66 and MS72, respectively. Ten leaves from each sample period were frozen in liquid nitrogen and ground into fine powder for total RNA extraction. An additional 20 leaves were used to measure chlorophyll and malondialdehyde (MDA) content.

2.2. Biochemical Indices

Chlorophyll content was measured according to the methods outlined by Lichtenthaler et al. [36]. A uniformly sized sample of pepper leaves was obtained using a punching machine, and 0.1 g of the sample was weighed and soaked in 20 mL of 95% ethanol under dark conditions for 36 h. Afterward, 100 uL of the supernatant was taken and placed in an enzyme-linked immunosorbent assay (ELISA) plate. The light absorption values were measured at 665 nm and 649 nm using an ELISA reader (ReadMax 1200, Shanpu, Shanghai, China). The chlorophyll a (Ca) and chlorophyll b (Cb) contents were calculated as follows: $Ca = 13.95A_{665nm} - 6.88A_{649nm}$, $Cb = 24.96A_{649nm} - 7.32A_{665nm}$. MDA content was estimated using thiobarbituric acid (TBA) [37]. Leaf material (1 g) was homogenized in liquid nitrogen and homogenized in 5 mL 0.1% (*w*/*v*) TCA (Shanghai Yuanye Bio-Technology,

Shanghai, China)solution. The homogenate was centrifuged at 10,000× g for 30 min (4 °C), and 0.5% TBA (Shanghai Yuanye Bio-Technology) was added to the obtained supernatant. The MDA concentration was measured spectrophotometrically at 532 nm. The analysis was performed in triplicate and displayed as mean ± SE. Statistical analysis was conducted using one-way ANOVA, and statistical significance ($p < 0.05$) was calculated using Duncan's test at the 5% level. Statistical analyses were performed using Sigmallot11 (Systat, San Jose, CA, USA) and Excel software 2016 (Microsoft, Redmond, WA, USA).

2.3. RNA Extraction, Sequencing, and Data Analysis

Total RNA was extracted using TRIzol reagent (Life Technologies, Carlsbad, CA, USA) according to the manufacturer's protocol. After extracting the total RNA, eukaryotic mRNA beads were enriched with oligo (dT). A NEBNext Ultra RNA Library Illumina Preparation Kit (NEB # 7530; New England Biolabs, Ipswich, MA, USA) was used for library preparation. The library was purified and amplified using polymerase chain reaction (PCR) after being connected to an Illumina sequencing adapter. The obtained cDNA library was sequenced by Gene Denovo Biotechnology Co. (Guangzhou, China) using an Illumina Novaseq 6000 (San Diego, CA, USA).

The raw sequence data reported in this study have been deposited in the Genome Sequence Archive (Genomics, Proteomics and Bioinformatics 2021) in the National Genomics Data Center (Nucleic Acids Res 2022), China National Center for Bioinformation/Beijing Institute of Genomics, Chinese Academy of Sciences (GSA: CRA018402), and are publicly accessible at https://ngdc.cncb.ac.cn/gsa (accessed on 16 August 2024) [38].

FASTP (version 0.18.0) [39] was used to filter raw data. Using Bowtie2, the ribosome reads were removed from the alignment, and the clean reads were compared with the chili genome solgenomics_ C. annuum _zunla [40]. The mapping readings for each sample were assembled using StringTie v1.3.1 [41] in a reference-based manner.

2.4. Differentially Expressed Gene Analysis and Weighted Gene Co-expression Network Analysis (WGCNA)

For each transcription region, RSEM [42] software was used to calculate the number of fragments read per kilobase transcript per million mappings (FPKM) to quantify its expression abundance and changes. Principal component analysis (PCA) and Pearson correlation coefficient analysis were performed on the gene expression levels of eighteen samples using OmicSmart (www.omicssmart. com) from Gene Denovo Biotechnology Co. (Guangzhou, China). DESeq2 (version 1.30.0) [43] was used to analyze the differential gene expression between the two groups. A false detection rate (FDR) < 0.05 and |log2 fold change (FC)| > 1 were used as screening criteria to obtain DEGs between the two samples. DEG sequences were compared with Gene Ontology (GO) databases (http://www.geneontology.org/, accessed on 10 March 2024) [44]. We also compared DEGs with the KEGG database using BLASTX and obtained the corresponding pathway annotation information [45].

Gene co-expression networks of DEGs for MS62 vs. MS72 at 3 DAI and 5 DAI were constructed using OmicSmart (http://www.omicsmart.com, accessed on 29 August 2024) [46]. Gene dendrograms were constructed using colors based on the correlations between the expression levels of genes, and were used to build clustering trees and divide the modules. In addition, correlations between the modules and samples were analyzed using WGCNA.

2.5. Validation of RNA-Seq Data by qRT-PCR

To verify the RNA-Seq results, nine genes were selected for qRT-PCR analysis using *CaActin* as an internal reference gene; specific primers were designed and are listed in Supplementary Table S1. The reactions were carried out in triplicate in a Vazyme FMR3 (Vazyme, Nanjing, China) system with AceQ Universal SYBR qPCR Master Mix (Vazyme). The reaction mixture consisted of 5 μL 2 × AceQ Universal SYBR qPCR Master Mix, 1.5 μL cDNA template, 0.4 μL each primer (10 μmol/μL), and 2.7 μL nuclease-free water. The amplification program was 95 °C for 30 s, followed by 40 cycles of 95 °C for 5 s and 60 °C for 30 s. Melting curve analyses were performed at the end of the 40 cycles (95 °C for 5 s followed by a constant increase from 60 °C to 95 °C). Relative gene expression levels were normalized to that of *CaActin* and determined with the $2^{-\Delta\Delta CT}$ method [47]. SigmaPlot v.11 (Systat, San Jose, CA, USA) software was used for statistical analysis and data display.

3. Results

3.1. Disease and Biochemical Indices Assessment

Morphological changes on the leaves of the two pepper cultivars inoculated with *V. dahliae* were recorded at 3 and 5 DAI (Figure 1A). On the inoculated leaves of MS66, yellowing surrounded the leaf veins at 3 DAI. No yellowing was observed in the leaves of MS72. After 5 DAI, all of the MS66 leaves turned yellow and began to wilt, while MS72 leaves showed small yellow areas, and no signs of wilt (Figure 1A). Compared with the resistant MS72, the chlorophyll a and b contents of the leaves of MS66 decreased rapidly (Figure 1B,C), reaching a reduction of 72.1% and 75.4% at 5 DAI, respectively. The MDA concentration in the susceptible cultivar MS66 increased rapidly (Figure 1D). During the tested inoculation periods, the MS72 leaves retained high chlorophyll content and exhibited low peroxidation of membrane lipids, supporting the resistance to the used *V. dahliae* strain.

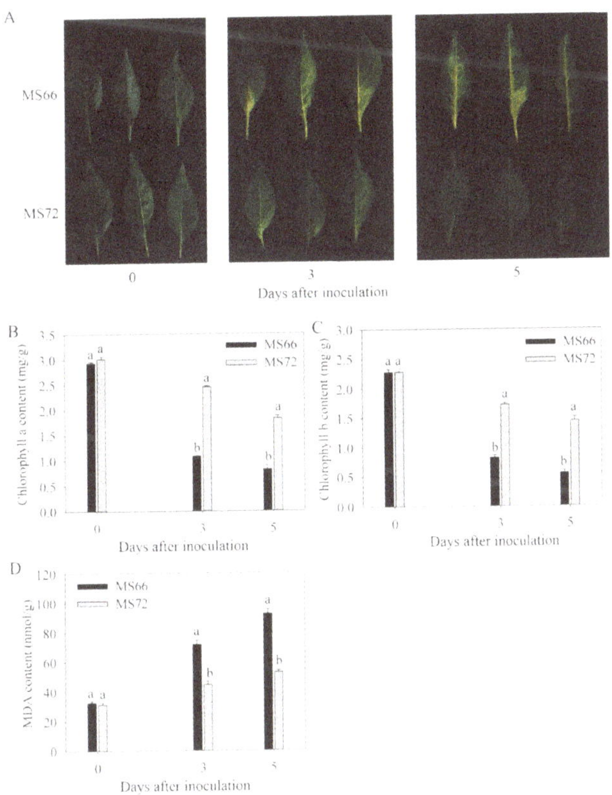

Figure 1. Interaction between *Verticillium dahliae* and MS66 (susceptible) and MS72 (resistant) pepper cultivars: (**A**) leaf phenotypes and symptom development; (**B**) chlorophyll a content; (**C**) chlorophyll b content; (**D**) malondialdehyde (MDA) content at 0, 3, and 5 days after inoculation (DAI). Histograms represent the mean (n = 3) ± SE. Values with the same letters are not significantly different according to Duncan's test ($p < 0.05$).

3.2. Sequencing Data Analysis

Through RNA sequencing, 111.9 billion raw reads were generated from 18 samples. After filtering, 111.0 billion high-quality clean reads were used for further analysis (Table 1). The average Q20 and Q30 values of clean data were 97.88% and 93.81%, respectively (Table 1). Additionally, the GC content of all samples exceeded 40% (Table 1). These results indicate that the sequencing quality is high and can be used for subsequent analysis. The PCA results show that three biological replicates can aggregate, forming six clear groups from 18 samples (Supplementary Figure S1). The Pearson correlation coefficient results showed a high correlation between biological replicates, which further demonstrated that the biological repeatability of sequencing data is high and scientific (Supplementary Figure S2). Collectively, the results show that the generated RNA-Seq sequencing data were reliable.

Table 1. Overview of the RNA sequencing data of 18 libraries. AF: After filter.

Sample	RawData (bp)	CleanData (bp)	AF_Q20 (%)	AF_Q30 (%)	AF_GC (%)
MS66-01 *	6,981,718,800	6,915,710,005	97.63%	93.34%	42.38%
MS66-02	6,258,249,000	6,197,996,540	97.49%	92.97%	42.63%
MS66-03	5,653,788,000	5,604,126,720	97.84%	93.71%	42.59%
MS66-31	5,674,388,700	5,629,439,592	98.05%	94.07%	42.13%
MS66-32	6,523,661,100	6,471,830,669	97.91%	93.82%	42.16%
MS66-33	6,125,935,800	6,079,521,125	98.06%	94.16%	42.21%
MS66-51	5,974,112,700	5,929,944,107	97.93%	93.83%	42.03%
MS66-52	6,492,510,300	6,429,441,859	98.05%	94.10%	42.01%
MS66-53	5,613,444,300	5,579,199,070	98.05%	94.11%	41.96%
MS72-01	5,662,599,600	5,608,834,245	97.56%	93.32%	42.72%
MS72-02	6,383,261,100	6,319,523,672	97.60%	93.37%	42.70%
MS72-03	6,431,895,000	6,377,583,237	97.71%	93.60%	42.70%
MS72-31	6,634,712,700	6,587,352,685	98.00%	94.00%	42.14%
MS72-32	6,406,942,800	6,364,254,446	98.11%	94.26%	42.22%
MS72-33	6,544,051,500	6,498,195,200	98.05%	94.20%	42.19%
MS72-51	6,177,454,500	6,125,703,488	97.94%	93.88%	42.09%
MS72-52	6,192,131,100	6,142,110,988	98.03%	94.11%	41.99%
MS72-53	6,225,490,200	6,168,686,984	97.89%	93.78%	41.89%

* The last digit of the sample name indicates biological duplication.

3.3. Transcriptome Profiles Analysis

3.3.1. DEGs of MS66 and MS72 Response to *V. dahliae* Inoculation

Upon inoculating the susceptible cultivar MS62 with *V. dahliae*, 5134 DEGs were identified at 3 DAI; 1916 of these genes were up-regulated, and 3218 down-regulated. At 5 DAI, 6909 genes were differentially expressed in MS66, with 2396 up-regulated and 4513 down-regulated. In the resistant cultivar MS72, 5073 genes were differentially expressed at 3 DAI, with 2204 up-regulated and 2869 down-regulated. After 5 days of inoculation, 4203 DEGs were identified in MS72, with 1848 up-regulated and 2335 down-regulated. Further analysis of the gene expression differences between two different cultivars after inoculation showed that, compared with susceptible cultivar MS62, 534 genes were differentially expressed in the resistant cultivar MS72 at 3 DAI, including 325 up-regulated and 209 down-regulated genes. At 5 DAI, a total of 2392 genes were differentially expressed in MS72 compared with MS66, with 1236 up-regulated and 1156 down-regulated genes (Figure 2).

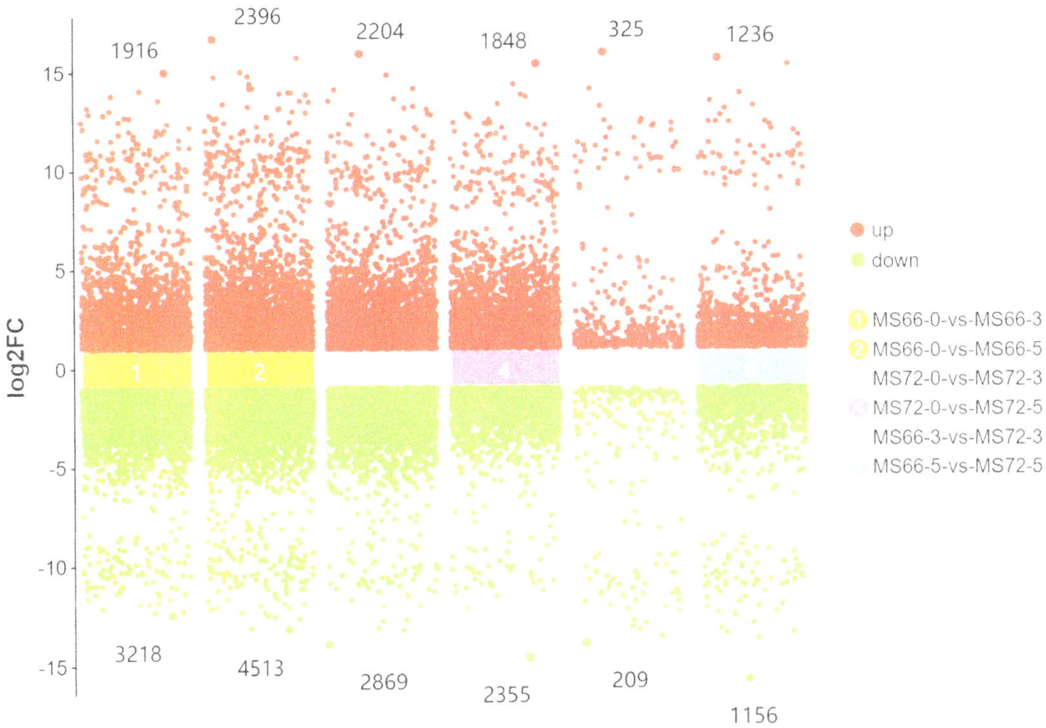

Figure 2. Number of differentially expressed genes (DEGs) in six comparative combinations of MS66 and MS72. Red dots indicate up-regulated DEGs, while green dots indicate down-regulated DEGs. The digits above the red dots and below the green dots represent the number of up-regulated DEGs and down-regulated DEGs, respectively.

3.3.2. GO Analyses of DEGs

The DEGs of the susceptible cultivar MS62 were mainly enriched in GO items related to chloroplast structure and kinases at 3 DAI, including photosynthetic membrane (GO: 0034357) and kinase activity (GO: 0004672) (Supplementary Figure S2, Table S2). At 5 DAI, the DEGs of MS62 were mainly enriched in GO items related to chloroplast structure (GO: 0009507) and photosynthesis (GO: 0015979) (Supplementary Figure S3, Table S2). Surprisingly, in the susceptible cultivar MS66, most DEGs in significantly enriched GO terms were down-regulated (Supplementary Figures S2 and S3).

In resistant cultivar MS72, the DEGs at 3 DAI were mainly enriched in GO items related to chloroplast structure and redox reactions, including photosynthetic membrane (GO:0034357), protein kinase activity (GO:0004672), and phosphorylation (GO:0006468) (Supplementary Figure S3, Table S2). At 5 DAI, the DEGs of MS72 were mainly enriched in GO items related to thylakoid (GO:0009579), photosynthetic membrane (GO:0034357), and cell wall organization (GO:0071555) (Supplementary Figure S3, Table S2).

At 3 DAI, the DEGs of MS72 compared with MS66 were significantly enriched in stress response and cell wall synthesis-related ontologies, such as the cell wall macromolecular metabolic process (GO:0044036) and response to oxidative stress (GO:0006979) (Figure 3A, Table S2). In contrast, GO terms significantly enriched for chloroplast structure and photosynthesis-related processes, including photosynthesis (GO:0015979) and photosynthetic membranes (GO:0034357), were observed in MS72 compared with MS66 at 5 DAI. In addition, compared with MS66, most DEGs of MS72 at 3 and 5 DAI were up-regulated in significantly enriched GO terms (Figure 3A, Table S2).

Figure 3. Top twenty enriched GO annotations (**A**) and KEGG annotations (**B**) for differentially expressed genes in MS66-3 vs. MS72-3 and MS66-5 vs. MS72-5 groups.

3.3.3. Pathway Enrichment Analyses of DEGs

In MS66-0 vs. MS66-3, MS66-0 vs. MS66-5, and MS72-0 vs. MS72-3 DEG comparisons, significantly enriched pathways were observed to be related to signaling pathways, including the MAPK signaling pathway in plants (ko04016), plant–pathogen interactions (ko04626), and plant hormone signal transduction (ko04075) (Supplementary Figure S3, Table S3). In the MS72-0 vs. MS72-5 comparison, the top 20 significantly enriched pathways were primarily related to metabolic pathways and photosynthesis, including phenylpropanoid biosynthesis (ko00940), stilbenoid, diarylheptanoid, and gingerol biosynthesis (ko00945), starch and sucrose metabolism (ko00500), and photosynthesis (ko00195) (Supplementary Figure S3, Table S3). Compared with the susceptible cultivar MS66, DEGs in the resistant cultivar MS72 at 3 DAI were significantly enriched in pathways such as cutin, suberin, and wax biosynthesis (ko00073) (Figure 3B, Table S2). At 5 DAI, DEGs in MS72 were significantly enriched in phenylpropanoid biosynthesis (ko00940), starch and sucrose metabolism (ko00500), photosynthesis (ko00195), and plant hormone signal transduction (ko04075) compared with MS66 (Figure 3B, Table S3).

3.3.4. Resistance Signaling Related Pathways Analysis to DEGs

We further analyzed the expression characteristics of signal pathways in DEGs. In the plant–pathogen interaction pathway, eight genes (Capana02g002878, Capana02g003359, Capana02g003630, Capana10g001171, Capana11g00029, Capana11g000407, Capana11g000420, Capana11g000436, and Capana11g000436) encoding calcium-binding proteins (CML), two *WRKY* transcription factors (Capana01g000167 and Capana09g001251), and mitogen-activated protein kinase 3-like (*MPK3*, Capana06g002562) in the susceptible cultivar MS66 were significantly up-regulated at 5 DAI. Conversely, in the resistant cultivar MS72, respiratory burst oxidase homologous protein B (*RbohB*, Capana08g001513), pathogenesis-related protein 1A-like (*PRB1*, Capana08g002191), and two LRR receptor-like serine/threonine protein kinases (*EFR*, MSTRG.19679 and MSTRG.4848) were up-regulated at 5 DAI (Figure 4A, Table S4).

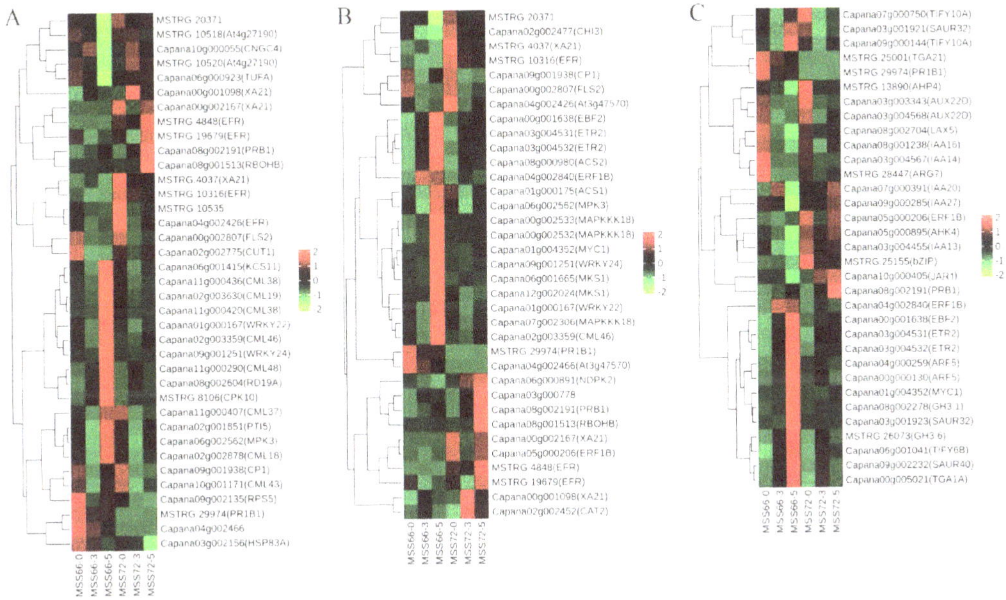

Figure 4. Heat map of DEGs in three resistance signaling-related pathways. (**A**) Plant–pathogen interaction; (**B**) MAPK signaling pathway—plant; (**C**) plant hormone signal transduction. The row Z-scores of the fragments per kilobase of transcript per million mapped fragments data are shown.

In the MAPK signaling pathway, three mitogen-activated protein kinase kinase kinase 2-like (Capana00g002532, Capana00g002533, and Capana07g002306), *MPK3* (Capana06g002562), two ethylene receptor 2 genes (*ETR2*, Capana03g004531 and Capana03g004532), two *WRKY* transcription factors (Capana01g000167 and Capana09g001251), and transcription factor MYC2-like (Capana01g004352) were significantly up-regulated at 5 DAI in the susceptible cultivar MS66. In contrast, in the resistant cultivar MS72, catalase isozyme 2 (Capana02g002452), *RbohB* (Capana08g001513), *PRB1* (Capana08g002191), and two *EFR* (MSTRG.19679 and MSTRG.4848) were up-regulated at 5 DAI (Figure 4B, Table S4).

Sixteen DEGs were up-regulated in the susceptible cultivar MS66 after inoculation; among these DEGs, two *CaETR2* (Capana03g004531 and Capana03g004532) and one ethylene-responsive transcription factor 1B-like (ERF1B, Capana04g002840) belong to the ethylene signaling pathway (Figure 4C, Table S4). In addition, seven DEGs belonged to the auxin signaling pathway (Figure 4C, Table S4), four DEGs, including two auxin-responsive protein (IAA, Capana07g000391, Capana09g000285) belonging to the auxin signaling pathway, jasmonic acid-amido synthetase JAR1-like (JAR1, Capana10g000405) belonging to the JA signaling pathway, and pathogenesis-related protein 1A-like (PR1, Capana08g002191) belonging to the SA signaling pathway, were up-regulated in the resistant cultivar MS72 after inoculation (Figure 4C, Table S4).

3.3.5. Key Gene Modules Screened by WGCNA

A gene clustering tree was constructed based on the correlation between gene expression levels, and the gene modules were divided according to the clustering relationship between the genes, yielding 16 stable expression modules (Figure 5A,B). Among these modules, the blue module contained the most genes, with 523 genes, whereas the grey module contained only two genes (Figure 5B).

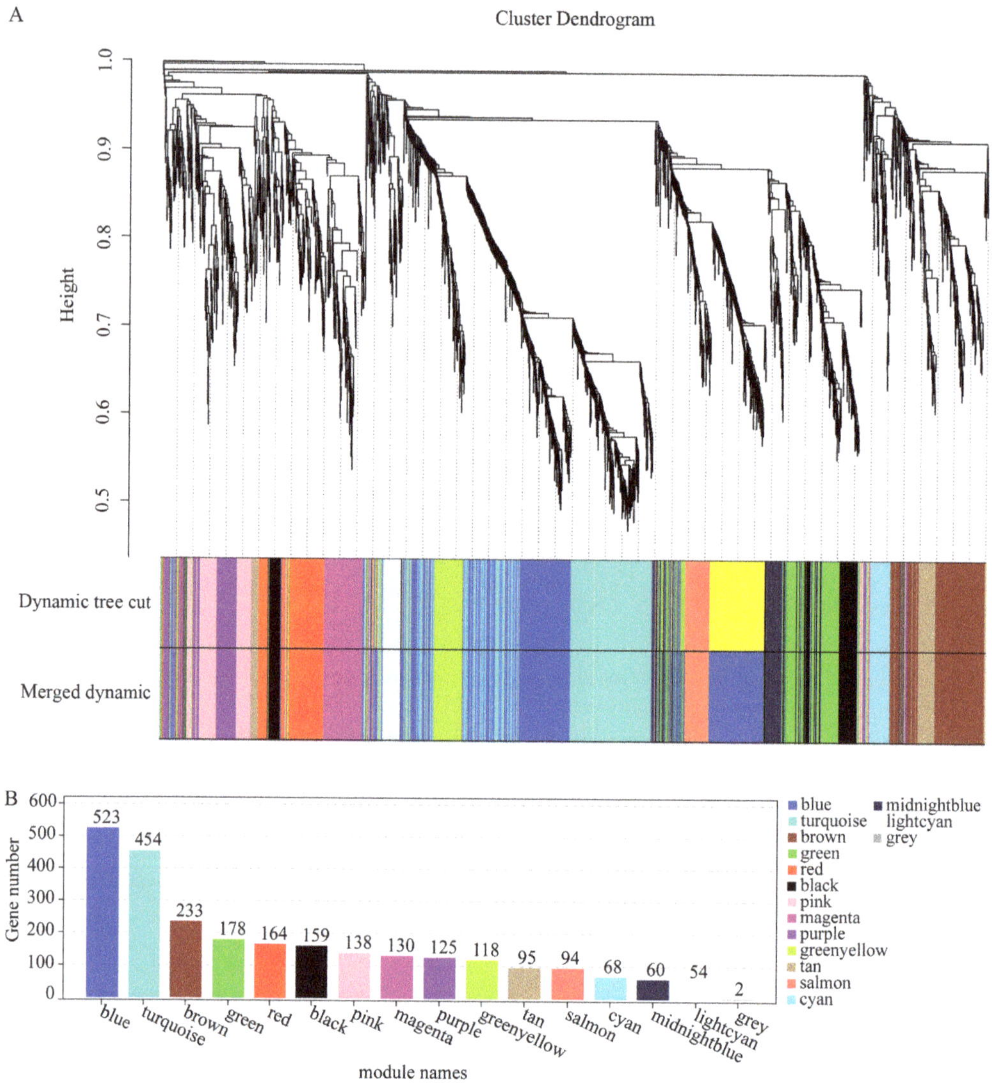

Figure 5. Key gene modules screened by WGCNA. (**A**) Module-level clustering diagram of DEGs based on WGCNA analysis. "Dynamic Tree Cut" refers to the module division based on clustering results. "Merged Dynamic" represents the division of merged modules with expression patterns similar to module similarity. (**B**) Number of genes in each module, with modules named after colors.

The module-sample expression pattern heatmap showed that the genes in the pink and purple modules of the resistant cultivar MS72 were significantly up-regulated after inoculation, whereas the genes of the susceptible cultivar MS66 did not show significant changes after inoculation (Figure 6). This indicates that these up-regulated genes may be involved in regulating disease resistance.

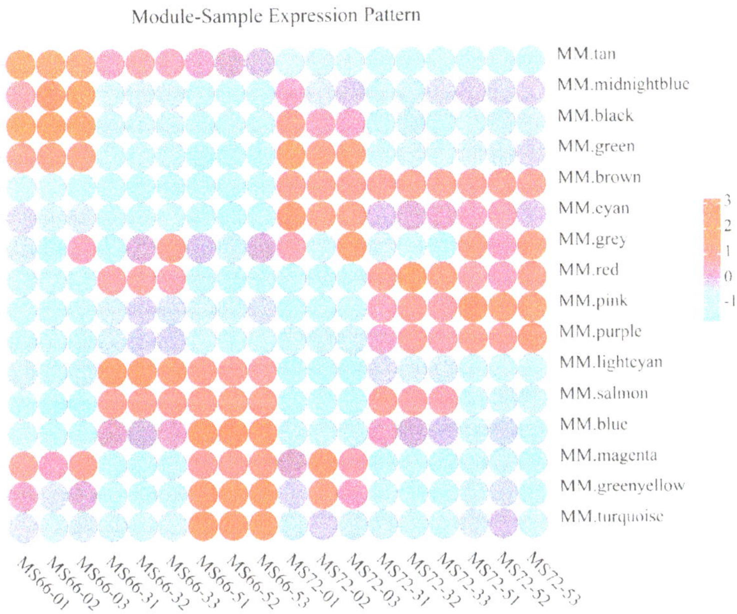

Figure 6. Expression pattern of each gene module in different samples. The *x*-axis represents different samples, and the *y*-axis represents different modules. The red color in the figure indicates high expression level, while the blue color indicates low expression level.

3.3.6. Hub Gene Analysis of Key Modules by WGCNA

In the pink module, six genes were located in the core position based on gene connectivity values and were selected as hub genes (Figure 7A, Table S5), including zeatin O-glucosyltransferase-like isoform X1 (MSTRG.32287), UPF0481 protein At3g02645 (Capana05g000620), GTP-binding protein-like (Capana09g002310), hypothetical protein BC332_17964 (Capana07g000070), wound-induced proteinase inhibitor 2-like (Capana03g001467), and Pin-II type proteinase inhibitor 7 (Capana03g001469). Among them, wound-induced protein inhibitor 2-like (Capana03g001467) and Pin-II type protein inhibitor 7 (Capana03g001469) both belong to the PIN-II protein family and exhibit fungicide-like effects [48].

In the purple module, 29 genes were located in the core position and were selected as hub genes (Figure 7B). These hub genes included transcription factor bHLH93-like (*bHLH93*, Capana04g000815). In addition, many of these hub genes are involved in plant resistance regulation, such as flower-specific defensin-like (*D1*, MSTRG.5910, MSTRG.24407, MSTRG.36069, and MSTRG.5904), serine protease inhibitor 1-like (*SPI*, Capana03g001482), miraculin-like (*MLP*, Capana06g000841 and Capana10g002167), chitin-binding lectin 1-like (*CBLs*, Capana03g000778), pectinesterase 2.2 (*PME2.2*, MSTRG.2104), and cellulose synthase-like protein G1 (*CSLG1*, Capana07g001101) (Figure 7B, Table S5).

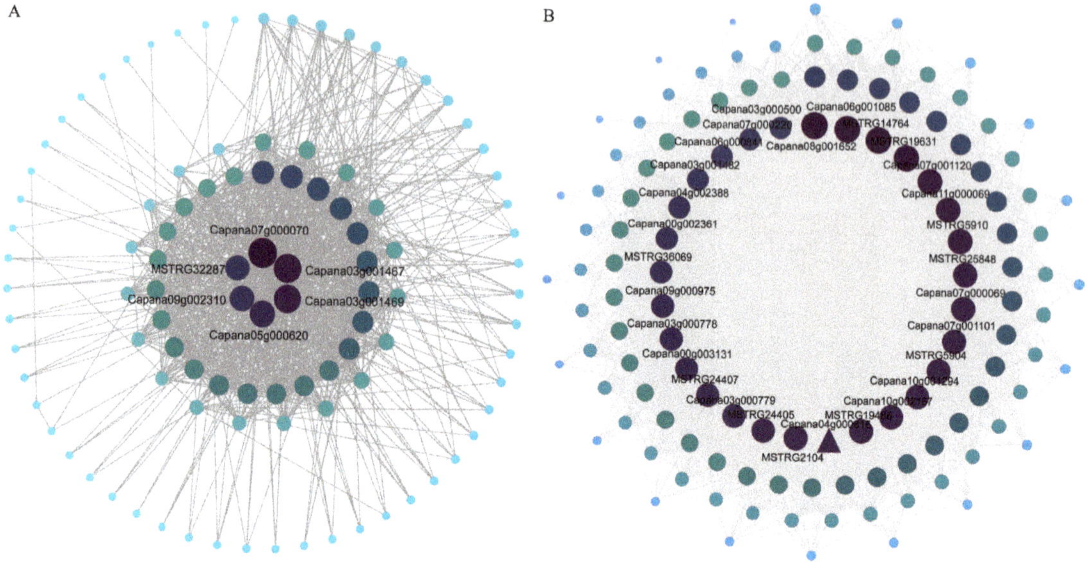

Figure 7. Gene interaction network analysis in the pink (**A**) and purple (**B**) modules. Different colors display connectivity abundance, with purple indicating high connectivity, blue indicating low connectivity, and triangles representing transcription factors.

3.4. Validation of RNA-Seq Data by Quantitative Real-Time (qRT)-PCR

The qRT-PCR and RNA-Seq results of nine hub genes showed similar expression profiles, with a Pearson correlation coefficient of 0.9281 ($p < 0.0001$) between the RNA-Seq and qRT-PCR data, confirming the reliability of the RNA-Seq data (Figure 8).

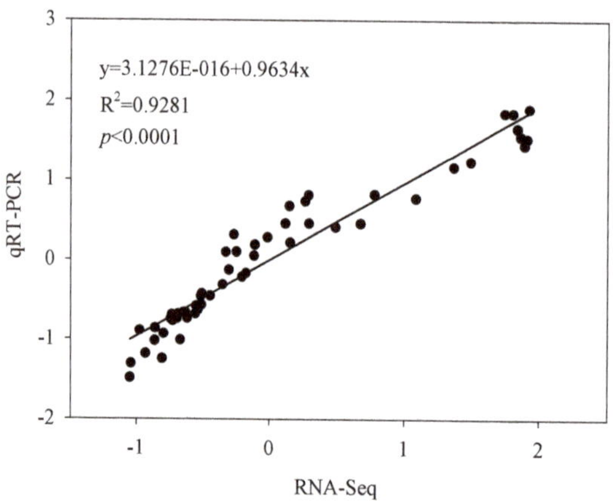

Figure 8. Validation of the RNA-Seq gene expression data by quantitative reverse transcription PCR (qRT-PCR). Nine hub genes from the RNA-Seq data were used for qRT-PCR assays. The row Z-scores of the fragments per kilobase of transcript per million mapped fragments and qRT-PCR data are shown.

4. Discussion

Pepper is an important global crop, serving not only as a key vegetable, but also as a vital industrial raw material for the pharmaceutical, agricultural, and cosmetics industries [2,3]. Verticillium wilt is a serious disease that causes yellowing of pepper leaves and significantly decreases pepper yield [8,13]. Although research has been conducted in recent years on pepper resistance to Verticillium wilt [26–28], and resistant cultivars have been developed [24,25], the gene regulatory network underlying the molecular mechanisms that confer resistance has not been explored. Furthermore, the key resistance genes and their regulatory mechanisms remain unclear, which is not conducive to breeding resistant pepper cultivars. Numerous studies have shown that the differential expression and regulation of resistance-related genes play important roles in plant responses to various pathogens [29–32]. Our research demonstrates that infection with *V. dahliae* causes lipid peroxidation of cell membranes in MS66, leading to a rapid decrease in chlorophyll content, whereas MS72 maintains high chlorophyll levels and low MDA content, indicating its resistance to *V. dahliae*. Subsequently, we used RNA-Seq to analyze the transcriptomes of resistant (MS72) and susceptible (MS66) cultivars at different time points after inoculation with *V. dahliae*. In MS66, DEGs at 3 DAI and 5 DAI were mainly enriched in GO terms related to chloroplast structure, kinase signaling, and photosynthesis, with most being down-regulated. In the resistant cultivar MS72, DEGs were mainly enriched in the chloroplast structure, kinase signaling, cell wall components, and antioxidant enzymes after inoculation. Further analysis of gene expression differences between the two cultivars at the same time points showed that DEGs were mainly enriched in the GO terms related to cell wall synthesis, reactive oxygen species stress, and photosynthesis, with most of these genes up-regulated in the resistant cultivar MS72. The GO analysis results of DEGs are consistent with the observed physiological changes, indicating that these genes are key genes in response to *V. dahliae* infection in different resistant peppers.

KEGG results showed that the DEGs in both cultivars were enriched in three signaling pathways: plant–pathogen interaction, MAPK signaling, and plant hormone signaling transduction. These findings align with those of previous research studies, which showed that inoculation with *V. dahliae* in plants such as cotton [49] and tomato [50] can lead to differential expression in three signaling pathways. There were significant differences in the expression patterns of the three signaling pathway genes between the two resistant pepper cultivars. Ca^{2+} influx is an early key response in PAMP signal transduction mediated by the channel protein CNGC [51]. Ca^{2+} can activates CDPKs, which are intracellular calcium signaling response proteins that play an important role in plant responses to abiotic stress and pathogens [52,53]. In this study, we found that *CaCNGC* and *CaCPK* expressions were up-regulated in the resistant cultivar MS72 compared to MS66. Simultaneously, the expression of the downstream respiratory burst oxidase homologous protein B (*CaRbohB*) was up-regulated. Mutating respiratory burst oxidase homologs in rice can reduce resistance to leaf blight in mutant rice [54], whereas transient expression of tomato *SlRbohB* in tobacco enhances resistance to *Botrytis cinerea* [55]. This suggests that the up-regulation of *CaRbohB* expression may be related to the resistance of MS72 to *V. dahliae*. *WRKY* is one of the largest transcription factor families in plants, widely involved in plant responses to biotic, abiotic, and hormonal stress. Among them, *WRKY22* and *WRKY24* are key transcription factors in plant–pathogen interactions and MAPK signaling pathways that can induce downstream signaling pathways to promote defense responses against pathogens and pests [56,57]. Interestingly, *CaWRKY22* and *CaWRKY24* expression were down-regulated in the resistant cultivar MS72. Research on sweet oranges has shown that *CsWRKY22* negatively regulates resistance to ulcer disease, and its overexpression promotes susceptibility to ulcer disease in Wanjincheng orange, whereas RNAi-mediated inhibition of *CsWRKY22* expression enhances resistance to ulcer disease [58].

Plant hormones play important regulatory roles in plant defense responses [59]. In this study, we found that most DEGs in the ethylene signaling pathway of the susceptible cultivar MS66 were significantly up-regulated, indicating that ethylene signaling may be involved in the regulation of yellowing and senescence of MS66 pepper leaves (Figure 4C, Table S4). Two *CaIAA*, one *CaJAR1*, and one *CaPR1* were up-regulated in the resistant cultivar MS72 after inoculation (Figure 4C, Table S4). IAA proteins play an important role in the auxin signaling pathway and are widely involved in plant development and responses to abiotic and biotic stresses. Notably, the overexpression of MeAux/IAA in tobacco can enhance resistance to *Xanthomonas axonopodis pv. manihotis* (Xan), whereas silencing MeAux/IAA confers sensitivity [60]. JA signal-deficient mutant *jar1* was found to be more susceptible to *Plasmodiophora brassicae* in *Arabidopsis* [61]. In cotton, the overexpression of *GhMYB36* can up-regulate the expression of *GhPR1*, thereby enhancing *V. dahlia* resistance. The down-regulation of *GhPR1* gene expression through VIGS makes plants more susceptible to *V. dahlia* infection and exhibits a more severe wilting phenotype compared to control plants [62]. This result indicates that the *CaIAA*, one *CaJAR1*, and one *CaPR1* may be involved in regulating resistance to *V. dahliae* of pepper.

Specific plant proteins are often activated under pathological conditions and are crucial for combating diseases. We used WGCNA to screen 35 hub genes in the pink and purple modules, including five *CaD1* genes (MSTRG.5910, MSTRG.5904, MSTRG.24405, MSTRG.24407, and MSTRG.36069) encoding plant defensins. Plant defensins are small alkaline peptides that function as protein translation inhibitors, alpha-amylase and protease inhibitors, or ion channel blockers [63,64]. Their most significant activity is their ability to inhibit the growth of various fungi and yeasts [63–65]. We also identified two *CaMLP* genes whose expressions were up-regulated in MS72 through WGCNA. MLPs are members of the Kunitz trypsin inhibitor family that play a role in protecting plants from pathogen invasion [66,67]. Overexpression of *StMLP1* in potatoes enhances their resistance to *Pseudomonas aeruginosa* [68]. Many plants and animals have CBLs, which can interfere with the synthesis and/or deposition of chitin in fungal cell walls. Therefore, most CBL compounds exhibit antifungal activity against plant pathogens [69–71], with the cell wall being the first line of defense against external pathogens [72]. A *CaCBL* (Capana03g000778) was screened as a hub gene and significantly up-regulated in MS72, indicating its potential involvement in resistance to *V. dahliae*. GO analysis revealed that the DEGs of the disease-resistant cultivar MS72 were enriched in cell wall-associated ontology. We further screened three hub genes, *CaCSLG1*, *CaPME2.2*, and *CaDIR21*, which are involved in cellulose modification and lignin accumulation, similar to known genes such as *HvCslD2* and *TaDIR13*. Barley *HvCslD2* encodes a specific cellulose-modifying enzyme, and silencing *HvSlD2* significantly reduced pathogen-induced cellulose deposition in the papillae, which increases susceptibility to powdery mildew [73]. Overexpression of the wheat *TaDIR13* in tobacco increases the total lignin accumulation, and exhibits strong resistance to *Phytophthora syringae* and *P. parasitica* [74]. The hub gene *CabHLH93* in resistant cultivars was significantly up-regulated after inoculation with *V. dahliae*. The bHLH superfamily is a transcription factor family in the helix loop helix (HLH) region [75] and is involved in plant biotic and abiotic stress responses [76,77]. In tobacco, the bHLH transcription factor GhPAS1 confers resistance to cotton Verticillium wilt by regulating specific members of the ERF and genes involved in lignin biosynthesis and plant defense [77]. In tomatoes, the effector RipI of Fusarium wilt induces host defense responses by interacting with the *bHLH93* transcription factor, thereby enhancing resistance to bacterial wilt disease [78]. In summary, these hub genes may be key genes that confer resistance to Verticillium wilt in MS72. However, this study only investigated the disease resistance regulatory network of leaves, and it is still unclear whether the selected hub genes play a role in regulating resistance throughout the entire plant. We are still exploring the functional mechanism of the hub genes revealed in this study to elucidate the molecular basis of the observed resistance to Verticillium wilt in MS72 pepper plants and provide ideas for breeding new cultivars resistant to Verticillium wilt. In addition, the research results offer potential strategies for breeding Verticillium wilt-resistant pepper varieties, such as overexpression of the *CabHLH93* gene in pepper, which could be a key method for breeding resistant cultivars.

5. Conclusions

Our research results indicate that inoculation with *V. dahliae* leads to a rapid decrease in chlorophyll content and an increase in MDA content in the susceptible cultivar MS66, ultimately resulting in leaf yellowing and wilting, whereas the resistant cultivar MS72 maintains high chlorophyll content and exhibits low peroxidation of membrane lipids. RNA-Seq analysis showed that, compared with MS66, DEGs in MS72 were mainly enriched in the GO terms related to cell wall, antioxidant, and photosynthesis, as well as in pathways such as cutin, suberin, wax biosynthesis, phenylpropanoid biosynthesis, and photosynthesis. DEGs related to the ethylene signaling pathway were up-regulated in MS66, whereas two *CaIAA*, one *CaJAR1*, and one *CaPR1* of plant hormone signaling pathways were up-regulated in MS72 after inoculation. Simultaneously, 35 hub genes were identified in up-regulated genes of MS72 through WGCNA. The key genes identified in this study may serve as a reference for improving pepper resistance through breeding programs. Ultimately, these findings may aid the development of robust disease-resistant crop cultivars that provide economic benefits and contribute to sustainable agriculture.

Supplementary Materials: The following supporting information can be downloaded at: https://www.mdpi.com/article/10.3390/horticulturae10111160/s1, Figure S1: Principal component analysis (PCA) of eighteen samples. Figure S2: Pearson correlation coefficient heatmap of eighteen samples. Figure S3: Top twenty significantly enriched GO terms of DEGs in MS66-0 vs. MS66-3, MS66-0 vs. MS66-5, MS72-0 vs. MS72-3, and MS72-0 vs. MS72-5 groups. Figure S4: DEGs enriched in KEGG pathways in MS66-0 vs. MS66-3, MS66-0 vs. MS66-5, MS72-0 vs. MS72-3, and MS72-0 vs. MS72-5 groups. Table S1: qRT-PCR primers used for validation of RNA-Seq data. Table S2: GO Information of DEGs. Table S3: KEGG pathway information of DEGs. Table S4: Information on three signaling pathways. Table S5: Hub genes of the pink and purple module.

Author Contributions: Conceptualization, X.H., Q.H. and L.H.; methodology, H.T., J.L. and L.H.; software, H.T., J.L. and X.H.; validation, Q.H., Q.Q., and Q.S.; formal analysis, H.T., J.L., L.H. and H.H.; investigation, L.H., H.T., J.L. and Q.S.; resources, Q.Q. and Q.S.; data curation, H.T., J.L. and X.H.; writing—original draft preparation, X.H., L.H., H.T., J.L., Q.Q. and Q.S.; writing—review and editing, Q.H., L.O. and H.H.; visualization, Q.H., L.O. and L.H.; supervision, L.O. and Q.H.; project administration, X.H., L.O., H.H. and Q.H.; funding acquisition, L.O., X.H. and H.H. All authors have read and agreed to the published version of the manuscript.

Funding: This study was funded by the Guangdong Basic and Applied Basic Research Foundation (2019A1515110599), the Maoming Science and Technology Plan Project (2019S001006), and the Science and Technology Tackle Key Problem of Guangdong Province (2021S0074).

Data Availability Statement: The raw sequence data reported in this paper have been deposited in the Genome Sequence Archive (Genomics, Proteomics and Bioinformatics 2021) at the National Genomics Data Center (Nucleic Acids Res 2022), China National Center for Bioinformation/Beijing Institute of Genomics, Chinese Academy of Sciences (GSA: CRA018402), and are publicly accessible at https://ngdc.cncb.ac.cn/gsa (accessed on 16 August 2024) [36]. The data are presented in the article and Supplementary Materials.

Acknowledgments: We thank Guangzhou Genedenovo Biotechnology Co. Ltd. for providing the methods for partial data analysis.

Conflicts of Interest: Author Qiucheng Qiu and Qidi Sun were employed by the company (Maoming Maoshu Seed Industry Technology Co., Ltd). The remaining authors declare that the research was conducted in the absence of any commercial or financial relationships that could be construed as a potential conflict of interest.

References

1. Yang, S.; Luo, X.; Jin, J.; Guo, Y.; Zhang, L.; Li, J.; Tong, S.; Luo, Y.; Li, T.; Chen, X.; et al. Key candidate genes for male sterility in peppers unveiled via transcriptomic and proteomic analyses. *Front. Plant Sci.* **2024**, *15*, 1334430. [CrossRef] [PubMed]
2. Parisi, M.; Alioto, D.; Tripodi, P. Overview of biotic stresses in pepper (*Capsicum* spp.): Sources of genetic resistance, molecular breeding and genomics. *Int. J. Mol. Sci.* **2020**, *21*, 2587. [CrossRef] [PubMed]
3. Baenas, N.; Belović, M.; Ilic, N.; Moreno, D.A.; García-Viguera, C. Industrial use of pepper (*Capsicum annum* L.) derived products: Technological benefits and biological advantages. *Food Chem.* **2019**, *274*, 872–885. [CrossRef] [PubMed]
4. Luo, X.-J.; Peng, J.; Li, Y.-J. Recent advances in the study on capsaicinoids and capsinoids. *Eur. J. Pharmacol.* **2011**, *650*, 1–7. [CrossRef] [PubMed]

5. Wang, Y.; Chen, B.; Cheng, C.; Fu, B.; Qi, M.; Du, H.; Geng, S.; Zhang, X. Comparative transcriptomics analysis reveals the differences in transcription between resistant and susceptible pepper (*Capsicum annuum* L.) varieties in response to anthracnose. *Plants* **2024**, *13*, 527. [CrossRef]
6. Ao, N.; Zou, H.; Li, J.; Shao, H.; Kageyama, K.; Feng, W. First report of Pythium aphanidermatum and Pythium myriotylum causing root rot on chili pepper (*Capsicum annuum* L.) in Guizhou, China. *Crop Prot.* **2024**, *181*, 106704. [CrossRef]
7. Evans, G.; McKeen, C.D. A strain of *Verticillium dahliae* pathogenic to sweet pepper in Southwestern Ontario. *Can. J. Plant Sci.* **1975**, *55*, 857–859. [CrossRef]
8. Goicoechea, N. Verticillium-induced wilt in pepper: Physiological disorders and perspectives for controlling the disease. *Plant Pathol. J.* **2006**, *5*, 258–265. [CrossRef]
9. Gurung, S.; Short, D.P.G.; Hu, X.; Sandoya, G.V.; Hayes, R.J.; Subbarao, K.V. Screening of wild and cultivated capsicum germplasm reveals new sources of Verticillium wilt resistance. *Plant Dis.* **2015**, *99*, 1404–1409. [CrossRef]
10. Hassan, O.; Ryu, H.; Choi, H.-W. First report of Verticillium wilt caused by *Verticillium dahliae* on chilli in South Korea. *Plant Dis.* **2023**. [CrossRef]
11. Grau, C.R. Verticillium wilt of alfalfa in Wisconsin. *Plant Dis.* **1981**, *65*, 843. [CrossRef]
12. Jing, R.; Li, H.; Hu, X.; Shang, W.; Shen, R.; Guo, C.; Guo, Q.; Subbarao, K.V. Verticillium wilt caused by *Verticillium dahliae* and *V. nonalfalfae* in potato in Northern China. *Plant Dis.* **2018**, *102*, 1958–1964. [CrossRef] [PubMed]
13. Bhat, R.G.; Smith, R.F.; Koike, S.T.; Wu, B.M.; Subbarao, K.V. Characterization of *Verticillium dahliae* isolates and wilt epidemics of pepper. *Plant Dis.* **2003**, *87*, 789–797. [CrossRef] [PubMed]
14. Lacasa, C.M.; Cantó-Tejero, M.; Martínez, V.; Lacasa, A.; Guirao, P. Performance of Aubergine rootstocks against *Verticillium dahliae* isolates in Southeastern Spain. *Agronomy* **2024**, *14*, 998. [CrossRef]
15. Xu, S.; Li, Y.Z.; Nan, Z.B. First report of Verticillium wilt of alfalfa caused by *Verticillium alfalfae* in China. *Plant Dis.* **2016**, *100*, 220. [CrossRef]
16. Joaquim, T.R. Reassessment of vegetative compatibility relationships among strains of *Verticillium dahliae* using nitrate-nonutilizing mutants. *Phytopathology* **1990**, *80*, 1160. [CrossRef]
17. Bardak, A.; Çelik, S.; Erdoğan, O.; Ekinci, R.; Dumlupinar, Z. Association mapping of Verticillium wilt disease in a worldwide collection of cotton (*Gossypium hirsutum* L.). *Plants* **2021**, *10*, 306. [CrossRef]
18. Wilhelm, S.; Wilhelm, S. Longevity of the Verticillium wilt fungus in the laboratory and field. *Phytopathology* **1955**, *45*, 180–181.
19. Goicoechea, N. To what extent are soil amendments useful to control Verticillium wilt? *Pest Manag. Sci.* **2009**, *65*, 831–839. [CrossRef]
20. Pei, D.; Zhang, Q.; Zhu, X.; Zhang, L. Biological control of Verticillium wilt and growth promotion in tomato by rhizospheric soil-derived *Bacillus amyloliquefaciens* Oj-2.16. *Pathogens* **2022**, *12*, 37. [CrossRef]
21. Ziazia, P.; Poulaki, E.G.; Gkizi, D.; Lozano, L.; Serrano, M.; Tjamos, S.E. Feeding the microbes: A strategy to control Verticillium wilt. *Agronomy* **2021**, *11*, 1946. [CrossRef]
22. Güneş, H.; Demir, S.; Durak, E.D.; Boyno, G. The effect of arbuscular mycorrhizal fungal species Funneliformis Mosseae and biochar against *Verticillium dahliae* in pepper plants under salt stress. *Eur. J. Plant Pathol.* **2024**. [CrossRef]
23. Poveda, J.; Calvo, J.; Barquero, M.; González-Andrés, F. Activation of sweet pepper defense responses by novel and known biocontrol agents of the genus *Bacillus* against *Botrytis cinerea* and *Verticillium dahliae*. *Eur. J. Plant Pathol.* **2022**, *164*, 507–524. [CrossRef]
24. Vasileva, K.; Todorova, V.; Masheva, S. Evaluation of collection of pepper (*Capsicum* spp.) resources for resistance to *Verticillium dahliae* Kleb. *Bulg. J. Agric. Sci.* **2019**, *25*, 1030–1038. [CrossRef]
25. González-Salán, M.M.; Bosland, P.W. Sources of resistance to Verticillium wilt in Capsicum. *Euphytica* **1991**, *59*, 49–53. [CrossRef]
26. Garmendia, I.; Aguirreolea, J.; Goicoechea, N. Defence-related enzymes in pepper roots during interactions with arbuscular mycorrhizal fungi and/or *Verticillium dahliae*. *BioControl* **2006**, *51*, 293–310. [CrossRef]
27. Coşkun, F.; Alptekin, Y.; Demir, S. Effects of arbuscular mycorrhizal fungi and salicylic acid on plant growth and the activity of antioxidative enzymes against wilt disease caused by *Verticillium dahliae* in pepper. *Eur. J. Plant Pathol.* **2023**, *165*, 163–177. [CrossRef]
28. Goicoechea, N.; Aguirreolea, J.; García-Mina, J.M. Alleviation of Verticillium wilt in pepper (*Capsicum annuum* L.) by using the organic amendment COA H of natural origin. *Sci. Hortic.* **2004**, *101*, 23–37. [CrossRef]
29. Zhao, Y.; Wang, D.; Ji, M.; Tian, J.; Ding, H.; Deng, Z. Transcriptome dynamic analysis reveals new candidate genes associated with resistance to Fusarium head blight in two Chinese contrasting wheat genotypes. *Int. J. Mol. Sci.* **2023**, *24*, 4222. [CrossRef]
30. Li, P.; Tan, X.; Wang, H.; Sun, L.; Jiang, J.; Fan, X.; Liu, C.; Zhang, Y. Transcriptome analysis of resistant and susceptible grapes reveals molecular mechanisms underlying resistance of white rot disease. *Hortic. Adv.* **2023**, *1*, 9. [CrossRef]
31. Zhang, Y.-Z.; Ma, Z.-B.; Li, W.-H.; Zhu, W.; Gao, S.-Q.; Zhao, Y.-H.; Liu, W. Transcriptome and metabolome profiling reveals key pathways and metabolites involved in defense against *Verticillium dahliae* in upland cotton. *Ind. Crops Prod.* **2023**, *196*, 116505. [CrossRef]
32. Feng, Z.; Wei, F.; Feng, H.; Zhang, Y.; Zhao, L.; Zhou, J.; Xie, J.; Jiang, D.; Zhu, H. Transcriptome analysis reveals the defense mechanism of cotton against *Verticillium dahliae* induced by hypovirulent fungus *Gibellulopsis Nigrescens* CEF08111. *Int. J. Mol. Sci.* **2023**, *24*, 1480. [CrossRef] [PubMed]

33. Sanogo, S.; Etarock, B.F.; Clary, M. Recovery of *Verticillium dahliae* from tall morningglory (*Ipomoea purpurea*) in New Mexico and its pathogenicity on chile pepper. *Plant Dis.* **2009**, *93*, 428. [CrossRef] [PubMed]
34. Zhu, X.; Soliman, A.; Islam, M.R.; Adam, L.R.; Daayf, F. *Verticillium dahliae's* isochorismatase hydrolase is a virulence factor that contributes to interference with potato's salicylate and jasmonate defense signaling. *Front. Plant Sci.* **2017**, *8*, 399. [CrossRef]
35. Zhu, X.; Sayari, M.; Islam, M.R.; Daayf, F. NOXA is important for *Verticillium Dahliae's* penetration ability and virulence. *J. Fungi* **2021**, *7*, 814. [CrossRef]
36. Lichtenthaler, H.K.; Wellburn, A.R. Determinations of total carotenoids and chlorophylls a and b of leaf extracts in different solvents. *Biochem. Soc. Trans.* **1983**, *11*, 591–592. [CrossRef]
37. Velikova, V.; Yordanov, I.; Edreva, A. Oxidative stress and some antioxidant systemsin acid rain-treated bean plants protective role of exogenous polyamines. *Plant Sci.* **2000**, *151*, 59–66. [CrossRef]
38. CNCB-NGDC Members and Partners. Database Resources of the National Genomics Data Center, China National Center for Bioinformation in 2022. *Nucleic Acids Res.* **2022**, *50*, D27–D38. [CrossRef]
39. Chen, S.; Zhou, Y.; Chen, Y.; Gu, J. Fastp: An ultra-fast all-in-one FASTQ preprocessor. *Bioinform. Oxf. Engl.* **2018**, *34*, i884–i890. [CrossRef]
40. Langmead, B.; Salzberg, S.L. Fast gapped-read alignment with Bowtie 2. *Nat. Methods* **2012**, *9*, 357–359. [CrossRef]
41. Pertea, M.; Pertea, G.M.; Antonescu, C.M.; Chang, T.-C.; Mendell, J.T.; Salzberg, S.L. StringTie enables improved reconstruction of a transcriptome from RNA-Seq reads. *Nat. Biotechnol.* **2015**, *33*, 290–295. [CrossRef] [PubMed]
42. Li, B.; Dewey, C.N. RSEM: Accurate transcript quantification from RNA-Seq data with or without a reference genome. *BMC Bioinform.* **2011**, *12*, 323. [CrossRef] [PubMed]
43. Love, M.I.; Huber, W.; Anders, S. Moderated estimation of fold change and dispersion for RNA-Seq data with DESeq2. *Genome Biol.* **2014**, *15*, 550. [CrossRef] [PubMed]
44. Ashburner, M.; Ball, C.A.; Blake, J.A.; Botstein, D.; Butler, H.; Cherry, J.M.; Davis, A.P.; Dolinski, K.; Dwight, S.S.; Eppig, J.T.; et al. Gene ontology: Tool for the unification of biology. *Nat. Genet.* **2000**, *25*, 25–29. [CrossRef]
45. Kanehisa, M.; Goto, S. KEGG: Kyoto Encyclopedia of Genes and Genomes. *Nucleic Acids Res.* **2000**, *28*, 27–30. [CrossRef]
46. Shi, W.; Wang, P.; Hu, R.; Wan, X.; Shen, H.; Li, H.; Wang, L.; Qiao, Y.; Jiang, G.; Cheng, J.; et al. Transcriptome analysis reveals hub genes in the hepatopancreas of Exopalaemon carinicauda in response to hypoxia and reoxygenation. *Aquac. Int.* **2021**, *29*, 1785–1811. [CrossRef]
47. Livak, K.J.; Schmittgen, T.D. Analysis of relative gene expression data using real-time quantitative PCR and the 2(-Delta Delta C(T)) method. *Methods* **2001**, *25*, 402–408. [CrossRef]
48. Rehman, S.; Aziz, E.; Akhtar, W.; Ilyas, M.; Mahmood, T. Structural and functional characteristics of plant proteinase inhibitor-II (PI-II) family. *Biotechnol. Lett.* **2017**, *39*, 647–666. [CrossRef]
49. Zhu, H.; Song, J.; Dhar, N.; Shan, Y.; Ma, X.-Y.; Wang, X.-L.; Chen, J.-Y.; Dai, X.-F.; Li, R.; Wang, Z.-S. Transcriptome analysis of a cotton cultivar provides insights into the differentially expressed genes underlying heightened resistance to the devastating Verticillium wilt. *Cells* **2021**, *10*, 2961. [CrossRef]
50. Tan, G.; Liu, K.; Kang, J.; Xu, K.; Zhang, Y.; Hu, L.; Zhang, J.; Li, C. Transcriptome analysis of the compatible interaction of tomato with *Verticillium dahliae* using RNA-sequencing. *Front. Plant Sci.* **2015**, *6*, 428. [CrossRef]
51. Clough, S.J.; Fengler, K.A.; Yu, I.-c.; Lippok, B.; Smith, R.K.; Bent, A.F. The Arabidopsis Dnd1 "Defense, No Death" gene encodes a mutated cyclic nucleotide-gated ion channel. *Proc. Natl. Acad. Sci. USA* **2000**, *97*, 9323–9328. [CrossRef] [PubMed]
52. Asano, T.; Hayashi, N.; Kobayashi, M.; Aoki, N.; Miyao, A.; Mitsuhara, I.; Ichikawa, H.; Komatsu, S.; Hirochika, H.; Kikuchi, S.; et al. A rice calcium-dependent protein kinase OsCPK12 oppositely modulates salt-stress tolerance and blast disease resistance. *Plant J. Cell Mol. Biol.* **2012**, *69*, 26–36. [CrossRef] [PubMed]
53. Dubiella, U.; Seybold, H.; Durian, G.; Komander, E.; Lassig, R.; Witte, C.-P.; Schulze, W.X.; Romeis, T. Calcium-dependent protein kinase/NADPH oxidase activation circuit is required for rapid defense signal propagation. *Proc. Natl. Acad. Sci. USA* **2013**, *110*, 8744–8749. [CrossRef] [PubMed]
54. Zhu, Y.; Su, H.; Liu, X.-X.; Sun, J.-F.; Xiang, L.; Liu, Y.-J.; Hu, Z.-W.; Xiong, X.-Y.; Yang, X.-M.; Bhutto, S.H.; et al. Identification of NADPH oxidase genes crucial for rice multiple disease resistance and yield traits. *Rice* **2024**, *17*, 1. [CrossRef]
55. Li, X.; Zhang, H.; Tian, L.; Huang, L.; Liu, S.; Li, D.; Song, F. Tomato SlRbohB, a member of the NADPH oxidase family, is required for disease resistance against Botrytis cinerea and tolerance to drought stress. *Front. Plant Sci.* **2015**, *6*, 463. [CrossRef]
56. Ramos, R.N.; Zhang, N.; Lauff, D.B.; Valenzuela-Riffo, F.; Figueroa, C.R.; Martin, G.B.; Pombo, M.A.; Rosli, H.G. Loss-of-function mutations in WRKY22 and WRKY25 impair stomatal-mediated immunity and PTI and ETI responses against *Pseudomonas syringae* pv. tomato. *Plant Mol. Biol.* **2023**, *112*, 161–177. [CrossRef]
57. Nie, W.; Liu, L.; Chen, Y.; Luo, M.; Feng, C.; Wang, C.; Yang, Z.; Du, C. Identification of the regulatory role of SlWRKYs in tomato defense against *Meloidogyne incognita*. *Plants* **2023**, *12*, 2416. [CrossRef]
58. Long, Q.; Du, M.; Long, J.; Xie, Y.; Zhang, J.; Xu, L.; He, Y.; Li, Q.; Chen, S.; Zou, X. Transcription factor WRKY22 regulates canker susceptibility in sweet orange (*Citrus sinensis* Osbeck) by enhancing cell enlargement and *CsLOB1* expression. *Hortic. Res.* **2021**, *8*, 50. [CrossRef]
59. Berens, M.L.; Berry, H.M.; Mine, A.; Argueso, C.T.; Tsuda, K. Evolution of hormone signaling networks in plant defense. *Annu. Rev. Phytopathol.* **2017**, *55*, 401–425. [CrossRef]

60. Fan, S.; Chang, Y.; Liu, G.; Shang, S.; Tian, L.; Shi, H. Molecular functional analysis of auxin/indole-3-acetic acid proteins (Aux/IAAs) in plant disease resistance in cassava. *Physiol. Plant.* **2020**, *168*, 88–97. [CrossRef]
61. Lemarié, S.; Robert-Seilaniantz, A.; Lariagon, C.; Lemoine, J.; Marnet, N.; Jubault, M.; Manzanares-Dauleux, M.J.; Gravot, A. Both the jasmonic acid and the salicylic acid pathways contribute to resistance to the biotrophic clubroot agent *Plasmodiophora brassicae* in *Arabidopsis*. *Plant Cell Physiol.* **2015**, *56*, 2158–2168. [CrossRef]
62. Liu, T.; Chen, T.; Kan, J.; Yao, Y.; Guo, D.; Yang, Y.; Ling, X.; Wang, J.; Zhang, B. The GhMYB36 transcription factor confers resistance to biotic and abiotic stress by enhancing *PR1* gene expression in plants. *Plant Biotechnol. J.* **2022**, *20*, 722–735. [CrossRef]
63. de Oliveira Carvalho, A.; Gomes, V.M. Plant Defensins-prospects for the biological functions and biotechnological properties. *Peptides* **2009**, *30*, 1007–1020. [CrossRef] [PubMed]
64. de Oliveira Carvalho, A.; Gomes, V.M. Plant defensins and defensin-like peptides-Biological activities and biotechnological applications. *Curr. Pharm. Des.* **2011**, *17*, 4270–4293. [CrossRef]
65. Yang, M.; Jiao, J.; Liu, Y.; Li, M.; Xia, Y.; Hou, F.; Huang, C.; Zhang, H.; Wang, M.; Shi, J.; et al. Genome-wide investigation of defensin genes in apple (*Malus* × *Domestica* Borkh.) and in vivo analyses reveal that *MdDEF25* confers resistance to *Fusarium Solani*. *J. Integr. Agric.* **2024**. [CrossRef]
66. Mondego, J.M.C.; Duarte, M.P.; Kiyota, E.; Martínez, L.; de Camargo, S.R.; De Caroli, F.P.; Alves, B.S.C.; Guerreiro, S.M.C.; Oliva, M.L.V.; Guerreiro-Filho, O.; et al. Molecular characterization of a miraculin-like gene differentially expressed during coffee development and coffee leaf miner infestation. *Planta* **2011**, *233*, 123–137. [CrossRef] [PubMed]
67. Ohkura, S.-I.; Hori, M.; Saitoh, K.; Okuzawa, T.; Okamoto, I.; Furukawa, N.; Shimizu-Ibuka, A. Structural and functional analysis of miraculin-like protein from Vitis vinifera. *Biochim. Biophys. Acta Proteins Proteom.* **2018**, *1866*, 1125–1130. [CrossRef]
68. Wang, B.; Wang, Y.; He, W.; Huang, M.; Yu, L.; Cheng, D.; Du, J.; Song, B.; Chen, H. StMLP1, as a Kunitz trypsin inhibitor, enhances potato resistance and specifically expresses in vascular bundles during Ralstonia solanacearum infection. *Plant J. Cell Mol. Biol.* **2023**, *116*, 1342–1354. [CrossRef]
69. Broekaert, W.F.; VAN Parijs, J.; Leyns, F.; Joos, H.; Peumans, W.J. A Chitin-binding lectin from stinging nettle rhizomes with antifungal properties. *Science* **1989**, *245*, 1100–1102. [CrossRef]
70. Trindade, M.B.; Lopes, J.L.S.; Soares-Costa, A.; Monteiro-Moreira, A.C.; Moreira, R.A.; Oliva, M.L.V.; Beltramini, L.M. Structural characterization of novel chitin-binding lectins from the genus artocarpus and their antifungal activity. *Biochim. Biophys. Acta* **2006**, *1764*, 146–152. [CrossRef]
71. Chen, C.-S.; Chen, C.-Y.; Ravinath, D.M.; Bungahot, A.; Cheng, C.-P.; You, R.-I. Functional characterization of chitin-binding lectin from *Solanum integrifolium* containing anti-fungal and insecticidal activities. *BMC Plant Biol.* **2018**, *18*, 3. [CrossRef] [PubMed]
72. Wolf, S. Cell Wall signaling in plant development and defense. *Annu. Rev. Plant Biol.* **2022**, *73*, 323–353. [CrossRef] [PubMed]
73. Douchkov, D.; Lueck, S.; Hensel, G.; Kumlehn, J.; Rajaraman, J.; Johrde, A.; Doblin, M.S.; Beahan, C.T.; Kopischke, M.; Fuchs, R.; et al. The barley (*Hordeum vulgare*) cellulose synthase-like D2 gene (*HvCslD2*) mediates penetration resistance to host-adapted and nonhost isolates of the powdery mildew fungus. *New Phytol.* **2016**, *212*, 421–433. [CrossRef] [PubMed]
74. Ma, Q.-H.; Liu, Y.-C. TaDIR13, a dirigent protein from wheat, promotes lignan biosynthesis and enhances pathogen resistance. *Plant Mol. Biol. Rep.* **2015**, *33*, 143–152. [CrossRef]
75. Toledo-Ortiz, G.; Huq, E.; Quail, P.H. The Arabidopsis basic/helix-loop-helix transcription factor family. *Plant Cell* **2003**, *15*, 1749–1770. [CrossRef]
76. Sun, X.; Wang, Y.; Sui, N. Transcriptional regulation of bHLH during plant response to stress. *Biochem. Biophys. Res. Commun.* **2018**, *503*, 397–401. [CrossRef]
77. Zhang, J.; Guo, M.; Wu, H.; Hou, L.; Li, S.; Chen, G.; Liu, L.; Liu, Z.; Lu, L.; Kabir, N.; et al. GhPAS1, a bHLH Transcription factor in upland cotton (*Gossypium hirsutum*), positively regulates *Verticillium dahlia* resistance. *Ind. Crops Prod.* **2023**, *192*, 116077. [CrossRef]
78. Zhuo, T.; Wang, X.; Chen, Z.; Cui, H.; Zeng, Y.; Chen, Y.; Fan, X.; Hu, X.; Zou, H. The *Ralstonia solanacearum* effector RipI induces a defence reaction by interacting with the bHLH93 transcription factor in *Nicotiana benthamiana*. *Mol. Plant Pathol.* **2020**, *21*, 999–1004. [CrossRef]

horticulturae

MDPI

Article

irus-Induced Silencing of a Sequence Coding for Loricrin-like Protein in *Phytophthora infestans* upon Infection of a Recombinant Vector Based on Tobacco Mosaic Virus

Rossella Labarile [1,*], Annamaria Mincuzzi [2], Roberta Spanò [2] and Tiziana Mascia [2,*]

[1] National Research Council (CNR), Institute of Chemical-Physical Processes, Via Amendola 165/A, 70126 Bari, Italy
[2] Department of Soil, Plant and Food Sciences, University of Bari "Aldo Moro", Via Amendola 165/A, 70126 Bari, Italy
* Correspondence: r.labarile@ba.ipcf.cnr.it (R.L.); tiziana.mascia@uniba.it (T.M.)

Abstract: *Phytophthora infestans* is the oomycete responsible for late blight disease of Solanaceae that causes both yield and economic losses. With the aim of reducing plant wilt and high management costs mainly due to wide fungicide applications, alternative eco-sustainable control strategies are needed. RNA interference (RNAi) is a powerful tool for gene function studies that can be accomplished by constitutive transformation or transient expression such as virus-induced gene silencing (VIGS) experiments. VIGS makes use of viruses to deliver sequences homologous to a target gene fragment and trigger RNAi. Indeed, a *P. infestans* ortholog of plant loricrin-like protein (*LLP*), named *PiLLP*, has been silenced using the direct infection of a recombinant vector based on the plant virus tobacco mosaic virus (TMV-*PiLLP*-1056), aiming to reduce the oomycete sexual reproduction. For this purpose, the gene coding for the green fluorescent protein (GFP) present in the TMV-GFP-1056 vector has been replaced with an antisense construct obtained by fusion PCR of the *PiLLP* 5′-UTR and 3′-UTR sequences. Here, we show that RNAi can be expressed in the A1 mating type of *P. infestans* strain 96.9.5.1 by VIGS using the direct infection of TMV-*PiLLP*-1056. We provide evidence that the recombinant vector can enter, replicate, and persist in mycelia of *P. infestans* where it induces the partial downregulation of the *PiLLP* transcript. Compared with the wild-type, the *PiLLP*-silenced A1 mating type had slower colony growth and a diminished virulence in detached tomato leaflets. This seems to be the first evidence of a constitutive gene downregulation of *P. infestans* using a recombinant vector based on a plus-sense RNA plant virus.

Keywords: A1; A2; mating type; *P. infestans*; recombinant virus; VIGS

1. Introduction

Oomycetes form a lineage of eukaryotic organisms that resemble filamentous fungi, but phylogenetic studies demonstrated a more extensive relatedness with brown algae and diatoms than with fungi, thus leading to the inclusion of oomycetes in the stramenopiles [1]. Similar to true fungi, oomycetes include threatening plant pathogens such as the heterotallic species *Phytophthora infestans* (Mont.) de Bary, the causal agent of potato late blight disease and other destructive diseases of Solanaceae [1]. Brownish black and water-soaked lesions covered by fungal sporulation are the main symptoms of late blight on leaves and stems, until plant wilt; similar symptoms are observed on other crops within the same family [2]. Late blight disease caused by *P. infestans* is well-known due to 1840s Irish potato famine, although it is considered a "re-emerging" disease [3] causing new severe epidemics that entail yield and economic losses and high management costs mainly due to up to ten sprays of fungicide applications per season [3].

With the aim of reducing pesticide applications and favoring disease control by natural pathogens and host resources, recently proposed control strategies exploit ways for the identification and inhibition of pathogen's effectors [4,5]. Upon being secreted into the plant host cytoplasm, effector proteins may transit to the host nucleus, where they modulate plant signaling and the plant's immune responses [6,7]. Indeed, host immunity can be negatively regulated favoring pathogen penetration or fulfilling its metabolic or structural needs during infection [8]. The plasticity of *P. infestans* genome and the abundance of predicted genes coding for disease-promoting effector proteins with high copy number variation, high substitution rates, and differential expression [9] pose continuous challenges to disease control. In *Phytophthora* spp., effectors drive the rapid sequence divergence of infection-related genes and have an exceptional high potential to adapt to new control strategies [10–12]. The situation was worsened by the recent emergence of more aggressive lineages of *P. infestans* such as the 13_A2 (Blue13) of the A2 mating type in Europe, which allowed the oomycete to exploit a sexual reproduction cycle. Until the late 1970s, only the mating type A1 of *P. infestans* was distributed worldwide and reproduced asexually allowing for some success in its management [13]. The emergence and worldwide distribution of the aggressive A2 mating types, enabled sexual cycles, induced genetic variations and gene transfer in populations of *P. infestans*, produced isolates with increased virulence, and created new challenges for late blight management [14]. Integrated pest management (IPM) strategies could provide more durable and environment friendly crop protection against late blight [15]. In recent years, research has been focused on exploiting photosynthetic purple non-sulfur bacteria (PNSB) for bioremediation [16,17] and biosensing in hybrid devices [18,19]. Their versatile metabolism allows them to be used as a promising inoculant to promote plant growth while controlling pathogenic attacks and enhancing the immune response of solanaceous plants [20,21]. RNA interference (RNAi) (syn. RNA silencing) is a promising and sustainable strategy that has been used to target and silence specific pathogens' genes by means of the small RNAs (sRNAs) produced during the RNAi process. Basically, there are two ways for RNAi exploitation in *Phytophthora* spp.: constitutive and transient gene silencing that have been proven to be effective [22]. Constitutive gene silencing implies that transformation constructs must be integrated in pathogen's genome and expressed via a specific promoter. In the transient gene silencing, pathogen's protoplasts are electroporated with in vitro-synthesized silencing constructs, mainly hairpin-like RNA structures, and then regenerated into new colonies. The recent host-induced gene silencing (HIGS) approach seems more effective than other strategies but implies that sRNA molecules, generated in transgenic plants expressing double-stranded RNAs (dsRNAs) of oomycete sequences, are then translocated into the pathogen involved in plant attack [23]. Similarly, plants infected with viral vectors carrying effectors' sequences, can induce synthesis of sRNAs and then target effector proteins [24]; this technique is named virus-induced gene silencing (VIGS). In both HIGS and VIGS, it is expected that sRNAs generated by plant RNAi can move into pathogen during infection process to silence the homologous endogenous gene. Being a more direct way, VIGS induces silencing via a recombinant viral vector triggering pathogen rather than host.

Recently, Cao et al. [25] demonstrated the nonpersistent acquisition of plant viruses by plant-associated fungi might commonly occur in nature, highlighting a possible role for fungi in the life cycle, spread, and evolution of plant viruses. Moreover, Mascia et al. [26,27] have shown that six taxonomically different plant viruses can replicate and persist in *P. infestans*, and, as in plants, a tobacco mosaic virus-based recombinant vector (TMV) triggers and suppresses RNAi in *P. infestans*. Furthermore, Mascia et al. [28] demonstrated that the same vector based on TMV could infect three fungal species within *Colletotrichum* genus, express a reporter gene ectopically, and silence the related reporter transgene expressed in *Colletotrichum* spp. Based on the above-mentioned evidence, it would not be surprising if an RNAi-based response could have the potential to be fruitfully exploited for specific gene silencing in *P. infestans*, using VIGS with a TMV recombinant vector. Recently, the CRISPR/Cas9 system-mediated gene knockout and in situ complementation of RNA-binding protein involved in oospore formation have been applied to oomycetes. For example, the PuM90, a stage-specific RNA-binding protein, which regulates oospore formation in *Pythium ultimum*, has been knocked out by the CRISPR/Cas9 approach. The resulting PuM90-knockout mutants were significantly defective in oospore formation [29]. In *P. infestans*, regulatory mechanisms involved in sexual reproduction are scarcely known except for a loricrin-like protein (LLP), which is strongly overexpressed during oospore formation [30].

PiLLP, a *P. infestans* ortholog of the plant LLP, was silenced by protoplast transformation. Compared with the wild-type (WT), the *PiLLP*-silenced transformants were defective in oospore formation, had slower colony expansion rates, produced less sporangia with lower germination and zoospore-release rates, and also had severely diminished virulence levels [31]. We selected *PiLLP* sequence as target to silence via the direct infection of a TMV-based recombinant vector carrying the 3′UTR and 5′UTR of the *PiLLP* coding sequence. The TMV RNA genome encodes four proteins of approx. 126, 183, 30, and 17.5 kDa. The 183 kDa polypeptide is synthesized by readthrough of the leaking termination codon of the gene for the 126 kDa polypeptide. Both the 126 and 183 kDa polypeptides, namely replicase (REP), are required for viral RNA replication. The 30 kDa polypeptide, also called movement protein (MP), is required for viral cell-to-cell movement whereas the 17.5 kDa polypeptide is the viral coat protein (CP) subunit. MP and CP are synthetized via a subgenomic RNA (sgRNA) strategy [32]. TMV-GFP-1056, one of the most used TMV-based vectors, drives the expression of the reporter green fluorescent protein (GFP) isolated from jellyfish, for functional studies in plants. It contains a duplicated sgRNA promoter for the CP synthesis from the heterologous tobamovirus tobacco mild green mosaic virus (TMGMV) to prevent homologous recombination between duplicated TMV CP sgRNA promoters and the CP sequence of the same virus [33–35]. The TMV sgRNA promoter directs the expression of any gene inserted downstream between the *Pac*I and *Xho*I cloning sites. For the purpose of this study, the gene coding for GFP in the TMV-GFP-1056 vector has been replaced with an antisense construct obtained by fusion PCR of the *PiLLP* 5′-UTR and 3′-UTR sequences. The resulting TMV-*PiLLP*-1056 vector did enter, replicate, and persist in mycelia of *P. infestans* where it induced the partial downregulation of the *PiLLP* transcript. Compared with the WT, the *PiLLP*-silenced A1 mating type strain 96.9.5.1 had slower colony growth and a diminished virulence in detached tomato leaflets. This seems to be the first evidence of a constitutive gene downregulation of *P. infestans* using a recombinant vector based on a plus-sense RNA plant virus.

2. Materials and Methods

2.1. *P. infestans* Mating Types and Culture Conditions

The mating types A1 (strain 96.9.5.1) and A2 (strain 13_A2) of *P. infestans* used in this study were kindly provided by Dr. David Cooke (The James Hutton Institute, Invergowrie, Dundee, Scotland). Both mating types were grown in a dark chamber on pea agar (PA) [36] medium at 18 ± 1 °C. To test the mycelial growth rate, culture plugs (5 mm diameter), cut from the actively growing edge of each colony, were transferred on PA Petri dishes, and the orthogonal colony diameters were measured after 14-day incubation. Sporangia of each mating type were aseptically collected from 20-day-old cultures and washed with 7 mL of distilled water, and the mycelia were gently rubbed with a glass rod. Zoospores were released after incubation of the sporangia suspension at 4 ± 1 °C for 2 h and then at room temperature for 1 h. To distinguish between the two mating types, each mycelium was grown in liquid culture, and the harvested material was filtered and air-dried on a sterile Whatman No.1 filter paper (Whatman, Maidstone, UK), before being freeze-dried at −20 ± 1 °C. The DNA was extracted from the mycelia with the DNeasy Plant Mini Kit (Qiagen, Hilden, Germany) after homogenization with Tissue Lyser II (Qiagen, Hilden, Germany), as manufacturer's recommendation. DNA preparations were subjected to PCR using primer pairs S1a/S1b [37] and PHYB-1/PHYB-2 [38] for mating type A1 and A2, respectively, and amplified with DreamTaq Hot Start DNA Polymerase (Thermo Fisher Scientific, Waltham, MA, USA). All PCRs were performed in a T100 Thermal Cycler (Bio-Rad, Bio-Rad Laboratories, Hercules, CA, USA). Annealing temperatures were 50 °C and 59 °C for S1a/S1b and PHYB-1/PHYB-2 primer pair, respectively. For further confirmation,

DNA preparations were amplified with W16-1/W16-2 primer pair [39,40] using 61 °C as primer annealing temperature. Then, PCR products were digested with *Hae*III (New England Biolabs, Ipswich, MA, USA). All PCR products were analyzed by 1.2% agarose gel electrophoresis in 1X TBE (90 mM Tris, 90 mM boric acid, 1 mM EDTA) and stained with GelRed® (Biotium, Biotium Inc., Fremont, CA, USA).

2.2. TMV-PiLLP-1056 Vector Construction

Total RNA was extracted from *P. infestans* mycelia with Eurogold RNA pure (Euroclone S.p.a., Pero, Milan, Italy) according to manufacturer's protocol. RNA was suspended in RNase-free sterilized water, and the concentration was adjusted to 1 µg/µL. Standard reverse-transcription reactions were carried out with High-Capacity cDNA Reverse Transcription Kit (Applied Biosystems™, Waltham, MA, USA). The 5′ and 3′ untranslated regions (UTR) of *PiLLP* were amplified in two separate reactions with specific primers (5′UTR for/rev and 3′UTR for/rev) annealed at 54 °C and 59 °C, respectively. Primers were designed to introduce a restriction site at 5′ for the *XhoI* enzyme and a restriction site at 3′ for the *PacI* enzyme to join the two PCR fragments. The size of 201 bp and 126 bp expected for the two amplified PCR products was estimated by electrophoresis in 2% low melting agarose gel in TBE buffer after GelRed® staining. Products were purified using Agarose Gel DNA Extraction Kit (Roche, Roche Diagnostics Corporation, Indianapolis, IN, USA) according to the manufacturer's instructions. Fusion PCR was carried out using Platinum™ Taq DNA Polymerase High Fidelity (Invitrogen™, Waltham, MA, USA) using 1:1 ratio and 5′UTR for/3′UTR rev primers annealed at 50 °C. The 327 bp amplified product was purified and eluted from agarose gel as described above. The fused fragment was cloned in pGem-T Easy Vector System (Promega, Promega Corporation, Madison, WI, USA) and inserted in anti-sense orientation into the *XhoI* and *PacI* sites of the pTMV-GFP-1056 vector [41]. To obtain the recombinant plasmid pTMV-*PiLLP*-1056, the plasmid pTMV-GFP-1056 was used as backbone. Briefly, both the fused fragment and the vector were double digested with *XhoI* and *PacI* (New England Biolabs, Ipswich, MA, USA), loaded on 2% low melting agarose gel in TBE buffer, stained with GelRed®, and eluted as described above. Ligation of the 5′UTR + 3′UTR joined fragments to the viral vector pTMV-1056 deprived of GFP was accomplished using T4 DNA ligase (Promega, Madison, WI, USA) followed by transformation of TOP10 *E. coli* competent cells (Invitrogen™, Waltham, MA, USA). The correct insertion of the construct in the pTMV-1056 plasmid was verified by DNA sequencing in both directions using GFP pro/CP pro primer pair. Vector map and cloning strategy are described in Figure S1. All primers used in this work are listed in Table 1.

2.3. Plant Inoculation

Inocula of TMV-*PiLLP* were prepared in tobacco (*Nicotiana tabacum* L.) cv. Samsun by rub-inoculation of a biologically active transcript synthesized from the *KpnI*-linearized pTMV-*PiLLP*-1056 plasmids with T7 RNA polymerase and the mMessage mMachine kit (Ambion, Thermo Fisher Scientific, Waltham, MA, USA), following the protocol of the manufacturer. To perform the assays, young plants were transplanted at cotyledon stage and used for the mechanical inoculation after two weeks. The second and third true leaves of four tobacco plants were inoculated with 30 µL of the transcription mixture obtained from pTMV-*PiLLP*-1056. Inocula of TMV-GFP-1056 were used as controls, whereas tobacco plants cv. Samsun inoculated with 100 mM (Na_2-K) phosphate buffer, pH 7.2 served as mock-inoculated controls. All the plants were grown at 22 ± 2 °C with a 16-hour light/8-hour dark photoperiod and daily monitored for symptom appearance. To ensure virus maintenance, systemically infected leaves of tobacco were collected at 10 days post inoculation (dpi), and the plant sap was extracted in 100 mM (Na_2-K) phosphate buffer, pH 7.2 for rub-inoculation onto a new stock of plants. Viral particles of TMV-*PiLLP*-1056 were purified from systemically infected tobacco leaves collected at 10 dpi as described by Lot et al. [42] and stored at −20 ± 1 °C in 50 mM NaCl, 30% glycerol solution until use. Viral RNA was extracted using phenol-chloroform protocol. Reverse-transcription polymerase chain reactions (RT-PCR) and DNA sequencing were used to assess the correct insertion of the 327 bp fused fragment into pTMV-1056 using GFPpro/CPpro primer pair. Furthermore, total RNA from infected tobacco leaves was extracted using RNeasy Mini Kit (Qiagen). Two µg of total RNA preparations were spotted onto positively charged nylon membranes (Roche) and preliminary tested for hybridization with digoxigenin (DIG)-labeled DNA probes specific for: (i) tomato mild green mosaic virus (TMGMV) coat protein (CP) present in TMV-1056 vector, (ii) GFP present in TMV-GFP-1056 vector, and (iii) 3′UTR + 5′UTR *PiLLP* fragment present in TMV-*PiLLP*-1056 vector. RNA preparations were separated by electrophoresis through 1.2% agarose gel in TBE buffer, stained with GelRed®, and blotted by capillarity onto positively charged nylon membranes. Northern blots were subjected to hybridization with digoxigenin (DIG)-labeled DNA probe specific for TMGMV CP.

Table 1. Primer pairs used for amplifications of the genes in this study.

Primer Name	Sequence	Size (bp)	Ta	Purpose	Ref.
S1a S1b	5′-AGGATTTCAACAA-3′ 5′-TGCTTCCTAAGG-3′	1250	50	Mating type identification	[37]
PHYB-1 PHYB-2	5′GATCGGATTAGTCAGACGAG-3′ 5′-GCGTCTGCAAGGCGCATTTT-3′	347	59	Mating type identification	[38]
W16-1 W16-2	5′-AACACGCACAAGGCATATAAATCTA-3′ 5′-GCGTAATGTAGCGTAACAGCTCTC-3′	557	61	Mating type identification	[39,40]
5′UTR for 5′UTR rev	5′-TCTCCTCGAGTCCTCCTCTCTGCGT-3′ 5′-CGACAATATTCCATAACTGCTAA GGTGGGGCTGTAA-3′	201	54	5′UTR region of *PiLLP* cloning	This study
3′UTR for 3′UTR rev	5′-TATGGAATATTGTCGAGT ACAGCAACGAGGCTTCGA-3′ 5′-CCCCCTTAATTAAACGACACTTGAAACATCGTGAG-3′	126	59	3′UTR region of *PiLLP* cloning	
5′UTR for 3′UTR rev	5′-TCTCCTCGAGTCCTCCTCTCTGCGT-3′ 5′-CCCCCTTAATTAAACGACACTTGAAACATCGTGAG-3′	327	50	5′UTR + 3′UTR of *PiLLP* fusion PCR	
GFP pro CP pro	5′-CGATGATGATTCGGAGGCTA-3′ 5′-ATCCAAGACACAACCCTTCG-3′	~500	55	pTMV-1056 plasmid sequencing	
REP for REP rev	5′-CAGTGCATATCGGCATTGTC-3′ 5′-CGTGACTCCTCTTCCGTCTC-3′	241	50	TMV replicase of *P. infestans* detection	
*Pi*EF for *Pi*EF rev	5′-ACTCCAAGAACGACCCTGCTAAGGCAACC-3′ 5′-TCCGACGGCTCGAGGATGACCATGCAG-3′	239	>60	Real Time PCR of ef1 of *P. infestans*	[31]
*Pi*ORF for *Pi*ORF rev	5′-CAGTCTCCCGTCACCAGTCC-3′ 5′-GGCAGTCGGCAATACCACAT-3′	~300	>60	Real Time PCR of *PiLLP*	

From left to right, each column shows primer name and related sequence, amplicon size (bp), annealing temperature (Ta), use purpose, and bibliographical reference (Ref.).

2.4. Infection in *P. infestans*

The infection of the *P. infestans* A1 and A2 mating types was performed with 50 µL of purified preparations of TMV-*PiLLP* (1 µg/mL) added to the sporangia growing in 100 mL of Pea Broth (PB) supplemented with 100 mg/mL ampicillin (Sigma-Aldrich, Merck KgaA, St. Louis, MO, USA). Fungal cultures inoculated only with 50 µL of 50 mM NaCl, 30% glycerol solution were used as negative control. All cultures were prepared in triplicates and incubated at 18 ± 1 °C under continuous shaking at 50 rpm. Total RNA, extracted from freeze-dried mycelia at 10 dpi, was reversely transcribed as reported above. RT-PCR was used to detect the TMV-*PiLLP* in mycelia of *P. infestans*. Reactions were carried out with REP for/rev primer pair using 50 °C as annealing temperature to amplify the replicase gene of TMV-1056. Purity and size of the amplified products was estimated as above. The comparative cycle threshold ($2^{-\Delta\Delta Ct}$) method corrected for PCR efficiencies (qPCR) was used to evaluate the relative abundance of *PiLLP* transcripts in the two mating types of *P. infestans* exposed to TMV-PiLLP-1056 inoculum. Reactions were carried out in an Applied Biosystems StepOne™ Real-Time PCR System apparatus using the Fast SYBR®Green Master Mix (Applied Biosystems, Waltham, MA, USA). Samples were run in triplicate, and reactions were set up in a total volume of 10 µL using 50 ng of cDNA and 250 nM each of for and rev primers pair. The constitutive elongation factor 1 (*ef1*) gene of *P. infestans* (PiEF), amplified with the PiEF for/rev primers pair, was used as housekeeping gene for *PiLLP* ORF gene transcripts normalization amplified by using the PiORF for/rev primer pair [31]. The cycling conditions were: 95 °C for 20 s, 40 cycles at 95 °C for 3 s, and 60 °C for 30 s. Specificity of the qPCR products was confirmed by analysis of the melting curves, which showed a single Tm peak obtained for each primer combination. Results were evaluated using the StepOne™ Software v2.3. PCR efficiency for each amplified fragment was derived from the slope of the regression line obtained by interpolating values from triplicates of five serial 1:2 dilutions of input cDNA amount and the relative Ct values using StepOne software (Applied Biosystems, Waltham, MA, USA). Means and standard deviations were calculated among three replicates. The whole experiment was repeated twice with a different set of biological samples.

2.5. Pathogenicity Test

Three-week-old *P. infestans* WT and TMV-*PiLLP*-1056 challenged mycelia cultured on PA in optimal growth conditions were used as inoculum for tomato leaf infection. For this purpose, fungal plates were flooded with 5 mL Pea Broth and scraped with a glass rod to release sporangia. Each resulting suspension was collected and used for the inoculation of detached leaflets of four-week-old tomato plants cv UC82. Leaflets were previously surface-sterilized with sodium hypochlorite solution (2% vol/vol), rinsed in sterile distilled water, aseptically air-dried, and transferred to sterile moist filter paper placed in Petri dishes (90 mm Ø). Each leaflet was inoculated on abaxial side with 30 µL of sporangia suspension (1×10^5 sporangia/mL) near the main vein and kept in growth chambers (16 ± 1 °C, 16 h light/8 h dark photoperiod). Each replicate was made of three leaflets placed in the same Petri dish, and two replicates were arranged for each tested inoculum. Disease severity was estimated by measuring the area of the lesions developed using ImageLab software 6.1 (Bio-Rad, Hercules, CA, USA).

3. Results and Discussion

3.1. *P. infestans* Mating Types Identification

The mating type of the two *P. infestans* isolates was confirmed by several tests. According to S1a/S1b PCR assay, A1 mating type was confirmed by the amplification of a product of about 1.25 kbp, whereas no amplification was observed for the A2 mating type with the same primer pair (Figure 1a). Amplification with PHYB-1/PHYB-2 primer pair aimed to target PHYB marker and resulted in a 347 bp fragment obtained only for the A2 mating type whereas no amplification product was obtained for the A1 mating type (Figure 1b). From both the *P. infestans* isolates, amplification with primer pair W16–1/W16–2 produced amplified products of 557 bp [39,40]. After digestion with *Hae*III endonuclease, two fragments of approx. 557 bp and 457 bp and one fragment of approx. 457 bp were observed, for the A1 and A2 mating types, respectively [39,40].

3.2. Infectivity in Tobacco Plants

To investigate whether *PiLLP* could be silenced by VIGS, the pTMV-*PiLLP*-1056 vector was assembled to express the 5′ and 3′ untranslated regions of *PiLLP* under the subgenomic RNA (sgRNA) promoter of TMV CP. The 201 bp amplified 5′UTR fragment and the 126 bp of 3′UTR fragment were joined by fusion PCR and the 327 bp resulting fragment was inserted in anti-sense orientation between the *XhoI* and *PacI* sites of pTMV-1056 in replacement of the ORF for GFP expression (Figure S1). The correct insertion and orientation of the 5′UTR + 3′UTR joined fragment were confirmed by sequencing the region of interest in both directions. Tobacco cv Samsun was used to biologically assess the successful transformation of the original TMV-GFP-1056 into TMV-*PiLLP*-1056 by the GFP screening, sequencing, and recombinant vector purification to validate the insertion of the *PiLLP* fragment and to prepare the inoculum, respectively. Thus, the recombinant transcript of TMV-*PiLLP*-1056 was synthesized in vitro, and its infectivity was tested in healthy plants of tobacco cv. Samsun. A transcript of TMV-GFP-1056 synthesized from the original pTMV-GFP-1056 was used as control, whereas tobacco plants cv. Samsun inoculated with 100 mM (Na_2-K) phosphate buffer, pH 7.2 served as mock-inoculated controls. Disease symptoms consisting in pale yellow discolorations along the main veins (Figure 2a) were observed on tobacco leaves at 7 dpi. These symptoms appeared also in the newly formed leaves and persisted up to 30 dpi. Possible contaminants of the original GFP sequence still present in the recombinant plasmid were excluded by the absence of fluorescence when plants were irradiated with a UV lamp (UVP, Upland, CA, USA) (Figure 2b), confirming the complete replacement of GFP with the 5′UTR + 3′UTR *PiLLP* fragment in the TMV-*PiLLP*-1056 vector. GFP fluorescence was clearly observed in plants rub-inoculated with the recombinant transcript TMV-GFP-1056 (Figure 2c), while mock-inoculated leaves did not show disease symptoms (Figure 2d) or GFP fluorescence when exposed to UV light (Figure 2e).

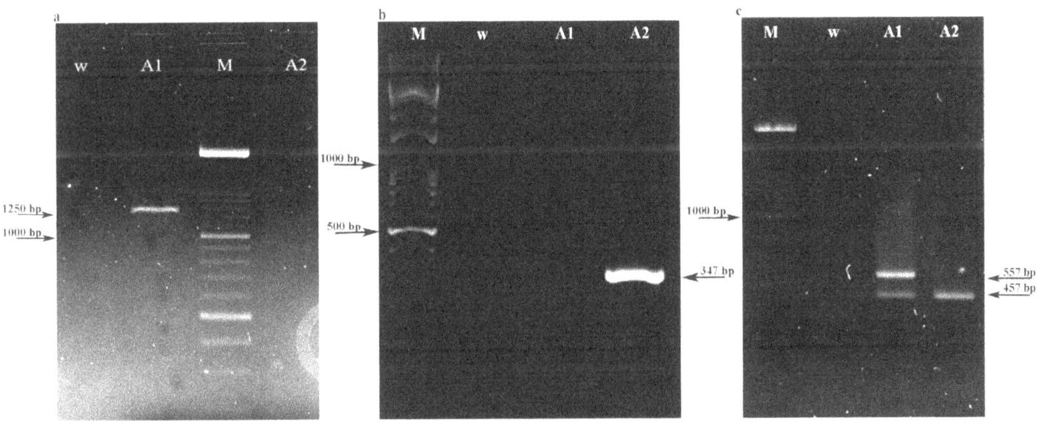

Figure 1. Detection of PCR products of *P. infestans* A1 and A2 mating type obtained with mating-type-specific primers. (**a**) product of 1250 bp (pointed by higher arrow) obtained from the A1 mating type with S1a/S1b primers; (**b**) product of 347 bp (pointed by lower arrow) obtained from the A2 mating type with PHYB-1/PHYB-2 primers; (**c**) products of 557 bp and 457 bp (pointed by two lower arrows) obtained from the A1 and A2 mating types with W16-1/W16-2 primers, followed by *Hae*III digestion. A1 = DNA extracted from WT A1 mating type; A2 = DNA extracted from WT A2 mating type; M = HyperLadder™ 100 bp (Bioline, Meridian Bioscience Inc., Cincinnati, OH, USA.); w = negative control (no DNA).

Replication of TMV-*PiLLP*-1056 in plants was confirmed by the detection of three sgRNAs corresponding to sequences replicated by the TMV MP promoter, TMV CP promoter, and TMGMV CP promoter, respectively (Figure 3). According to the TMV expression strategy and the pTMV-*PiLLP*-1056 map shown in Figure S1, sgRNA 1 has been synthesized by the TMV MP promoter; it has an estimated size of 2.4 kb and contains sequences of TMV-MP, 3′-5′ *PiLLP*, and TMGMV CP; sgRNA 2 has been synthesized by TMV CP promoter; it has an estimated size of 1.6 kb and contains sequences of 3′-5′ *PiLLP* and TMGMV CP; sgRNA3 has been synthesized by the TMGMV CP promoter; it has an estimated size 0.48 kb and contains the sequence of TMGMV CP. All sgRNAs share the sequence coding for the CP of TMGMV, which thus can be revealed by the 643 bp-long TMGMV CP probe. However, the three sequences are under the control of different promoters and the relative abundance of each sgRNA may vary according to the efficiency of its transcription, which was not evaluated.

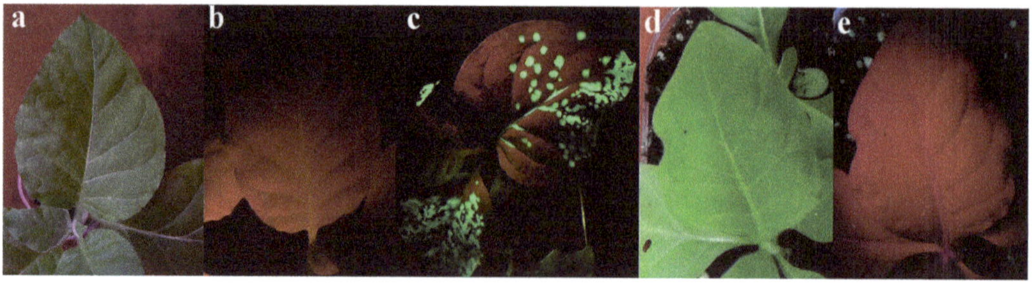

Figure 2. Exposure of tobacco plants to UV lamp to check the replacement of GFP with 5′UTR + 3′UTR *P. infestans* ortholog of plant loricrin-like protein (*PiLLP*) in TMV-*PiLLP*-1056 vector. (**a**) Disease symptoms consisting in pale yellow discolorations along the main veins in a tobacco leaf systemically infected with TMV-*PiLLP*-1056 not exposed to UV light. (**b**) No residual GFP fluorescence of tobacco leaf infected by TMV-*PiLLP*-1056 when exposed to UV light. (**c**) GFP fluorescence in tobacco plant inoculated with TMV-GFP-1056 used as control, (**d**) tobacco leaf of mock-inoculated plant used as negative control not exposed, and (**e**) exposed to UV light.

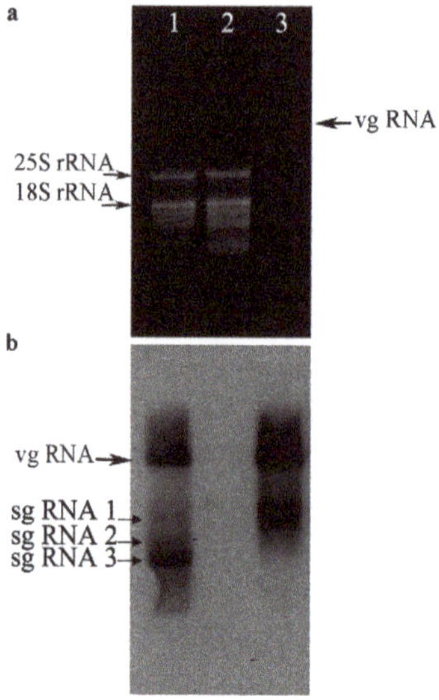

Figure 3. Detection of TMV-*PiLLP*-1056 RNA by northern blot hybridization with a 643-bp DIG-labeled DNA probe for TMGMV CP. (**a**) RNA gel electrophoresis and (**b**) northern blot hybridization of: lane 1 = 2 μg of total RNA preparation extracted from tobacco plant infected by TMV-*PiLLP*-1056 at 14 dpi; lane 2 = 2 μg of total RNA preparation extracted from mock-inoculated tobacco plant used as negative control; lane 3 = 0.2 μg of RNA preparation extracted from purified preparation of TMV-*PiLLP*-1056 used as control. Arrows indicate the positions of tobacco leaf 25S and 18S rRNAs of 3.7 and 1.9 kb, respectively, (**a**) and TMV genomic RNA (vgRNA) and subgenomic RNA1, RNA2, RNA3 (sgRNAs) RNAs (**b**).

Further confirmation of the presence of TMV-*PiLLP*-1056 and absence of any transcript expressing GFP in plants was obtained by dot blot hybridization of sap extracted from infected tobacco leaves collected at 14 dpi and hybridized with DIG-labeled DNA probes specific for the TMGMV CP, GFP, and 3′-5′ *PiLLP* fragment (Figure 4). Specific hybridization signals were observed in leaf infected with TMV-*PiLLP*-1056 and in a leaf back-inoculated with sap of a TMV-*PiLLP*-1056 infected leaf using TMGMV CP probe whereas the GFP expression was absent using GFP probe. Hybridization signals were observed in the same samples using 3′-5′ *PiLLP* probe, confirming the presence of the recombinant viral vector.

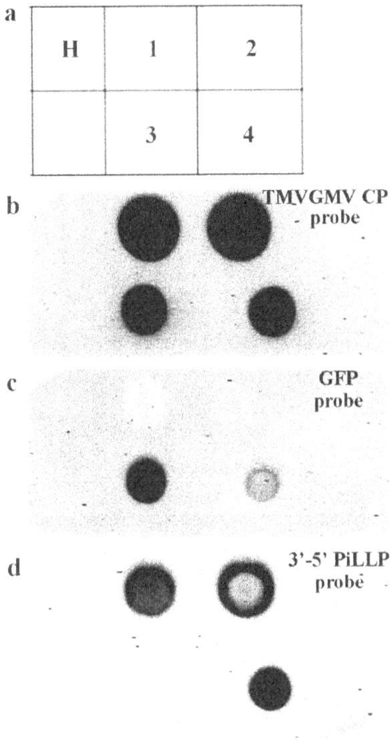

Figure 4. Detection of TMV-*PiLLP*-1056 presence in sap extracted from cv Samsun tobacco plants with DIG-labeled DNA specific probes. Dot blot hybridization scheme (**a**): H = sap extracted from healthy tobacco plant; 1 = sap extracted from leaf infected with TMV-*PiLLP*-1056 at 14 dpi; 2 = sap extracted from leaf back-inoculated with sap of a TMV-*PiLLP*-1056 infected leaf; 3 = 2 ng of recombinant plasmid pTMV-GFP-1056 served as positive control; 4 = 2 ng of recombinant plasmid pTMV-*PiLLP*-1056 used as positive control. Results of dot blot hybridization with (**b**) TMGMV CP, (**c**) GFP, and (**d**) 3′-5′ *PiLLP* probes. Non-specific hybridization signals are visible in spot 4 hybridized with GFP probe and in spot 3 hybridized with 3′-5′ *PiLLP* probe.

TMV-*PiLLP*-1056 infection was transferred to a larger number of tobacco plants by rub-inoculation with sap extracted from plants infected with the recombinant transcript at 14 dpi. TMV-*PiLLP*-1056 was purified from systemically infected leaves collected at 21 dpi. Virus preparations were viewed under the transmission Philips-Morgagni 282D electron microscope and compared to standard TMV particles; TMV-*PiLLP*-1056 particles appeared swollen at one end, probably as consequence of the TMGMV CP that replaced the authentic TMV CP in the recombinant vector. Particles shorter than the standard TMV modal length were also present (Figure 5).

3.3. *Infectivity in P. infestans*

A purified preparation of the recombinant virus TMV-*PiLLP*-1056 was used for the inoculation of liquid cultures of *P. infestans* A1 and A2 mating types. The presence of the recombinant virus in mycelia of *P. infestans* collected at 10 dpi was confirmed by RT-PCR targeting TMV-*PiLLP*-1056 replicase, which was not present in sgRNAs (see pTMV-*PiLLP*-1056 map in Figure S1). The PCR yielded the expected amplified product of 241 bp that was observed in both *P. infestans* mating types exposed to the inoculation of the TMV-*PiLLP*-

1056, while no bands were amplified in the mock-inoculated controls (Figure 6a). RT-qPCR with the primer pair targeting the *PiLLP* ORF showed a 3.5-fold and a seven-fold reduced expression of *PiLLP* ORF transcript in mating types A1 and A2, respectively, infected with TMV-*PiLLP*-1056 at 10 dpi (Figure 6b) compared to A1 and A2 mock-inoculated controls, used as reference sample for gene expression normalization. Elongation factor 1 alpha (*EF-1a*) was used as housekeeping gene for target genes normalization. Melting curve of the housekeeping gene *ef1* and *PiLLP* ORF after RT-qPCR showed a single peak, suggesting the absence of non-specific amplified products. Overall, these results suggest that the recombinant vector TMV-*PiLLP*-1056 was able to enter mycelia of both the mating types of *P. infestans* and induce silencing of the *PiLLP* gene due to sequence identity between the *PiLLP* UTRs regions in the recombinant vector and the constitutive gene transcribed in *P. infestans*.

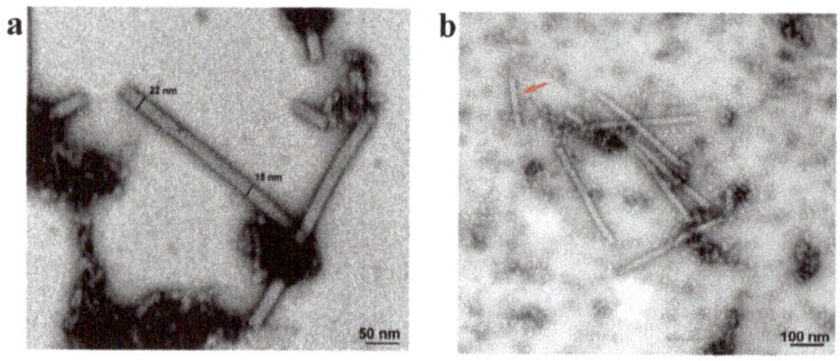

Figure 5. Electron microscope pictures of TMV-*PiLLP*-1056 and TMV virus preparations. TMV-*PiLLP*-1056 particles purified from Samsun tobacco plants showing an increase in particle diameter from 18 to 22 nm (**a**). In (**b**), arrow points particles shorter than the standard TMV modal length (Courtesy Dr. Angelo De Stradis IPSP CNR, Bari).

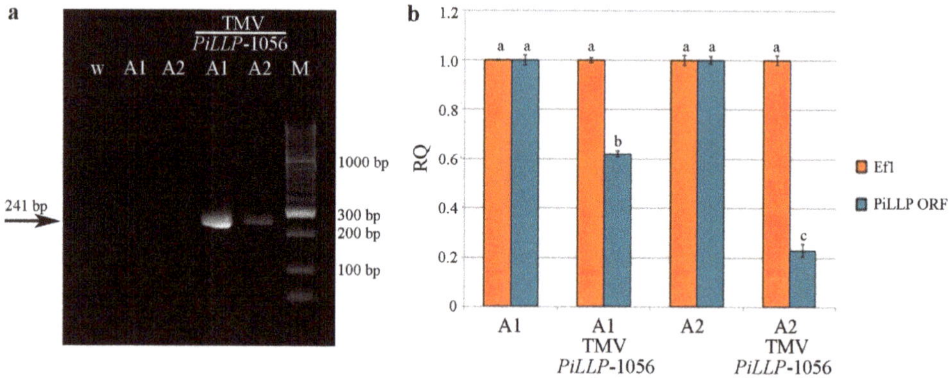

Figure 6. Amplification of a fragment of TMV-*PiLLP*-1056 replicase and *PiLLP* ORF in mycelia of *P. infestans* A1 and A2 mating types at 10 dpi with of TMV-*PiLLP*-1056. (**a**) PCR assay of TMV-*PiLLP*-1056 replicase yielded the expected amplified product of 241 bp (pointed by arrow) in infected mycelia compared to mock-inoculated controls; M = HyperLadder™ 100 bp (Bioline, Meridian Bioscience Inc., Cincinnati, OH, USA); w = negative control (no DNA); (**b**) Relative abundance of *PiLLP* ORF transcripts (blue bar) compared with that of the housekeeping gene *Ef1* (orange bar) in infected mycelia compared to mock-inoculated controls. RQ values ± St.dev. were calculated as mean of three biological replicates obtained from RT-qPCR. Letters indicate statistically significant differences ($p \leq 0.05$, Tukey post hoc test), for each genotype. The mock-inoculated condition was used as reference mycelia for gene expression normalization. Elongation factor 1 (*Ef1*) was used as housekeeping gene for target genes normalization.

Mycelia from liquid cultures of the two mating types exposed to TMV-*PiLLP*-1056 infection (denoted A1 TMV-*PiLLP* and A2 TMV-*PiLLP*), and the relative mock–inoculated controls were transferred to PA plates and grown under standard conditions. Compared with mock-inoculated control, the growth of A1 TMV-*PiLLP* was slower than A1 mock-inoculated control, whereas A2 TMV-*PiLLP* did not grow. The experiment was repeated twice with identical results.

Due to unavailability of the A2 TMV-*PiLLP*-1056, the variation in pathogenicity resulting from the partial downregulation of *PiLLP* transcript was estimated only for the A1 TMV-*PiLLP*. Infectivity/pathogenicity of A1 TMV-*PiLLP* was tested using detached tomato leaves instead of tobacco, since tomato is one of the most economically important hosts of *P. infestans*. Thus, detached tomato leaflets were inoculated in Petri dishes with equivalent amounts of germinated sporangia of infected A1 TMV-*PiLLP* and mock-inoculated A1 and A2 mating types as control. At 7 dpi, typical water-soaked lesions developed on leaflets inoculated with A1 and A2 mock controls, where the lesions observed in the A2 mating type were larger than in A1. Moreover, lesions induced by A1 TMV-*PiLLP* were smaller in size than those produced by the A1 mock controls and with poor aerial mycelium (Figure 7a,b). The colony diameter of A1 TMV-*PiLLP* cultures plated on PA was also smaller than that of the A1 WT.

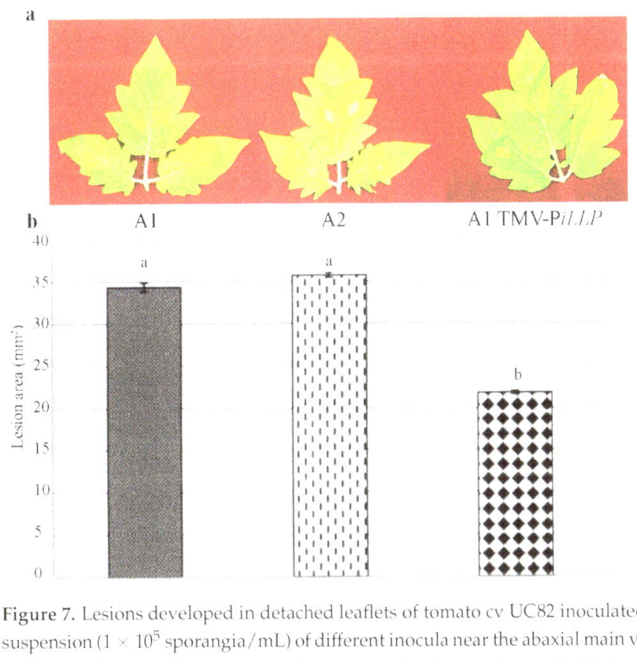

Figure 7. Lesions developed in detached leaflets of tomato cv UC82 inoculated with 30 μL sporangia suspension (1×10^5 sporangia/mL) of different inocula near the abaxial main vein. Lesions developed in detached leaflets of tomato cv UC82 inoculated with A1, A2 and A1 TMV-*PiLLP* (**a**). Sizes of lesions in mm^2 induced by A1 (grey bar), A2 (dotted bar) and A1 TMV-*PiLLP* (checkered bar) inocula (**b**). Vertical bars ± St.dev represent the mean of three biological replicates. Letters indicate statistically significant differences ($p \leq 0.05$, Tukey post hoc test), for each genotype. A1 = lesion area induced by germinated sporangia of the A1 mating type; A2 = lesion area induced by germinated sporangia of the A2 mating type; A1 TMV-*PiLLP* = lesion area induced by germinated sporangia of the A1 mating type infected with TMV-*PiLLP*-1056.

The same approach was used to evaluate the *PiLLP* downregulation effect on oospore development. WT isolates, infected A1 TMV-*PiLLP* and A2 TMV-*PiLLP* isolates, were mated on tomato leaflets with equivalent amounts of germinated sporangia for each isolate. Although the experiment was repeated twice, no oospore development was observed.

Probably, this might be due to scarce α1 hormone produced by A1 mating type that inhibits A2 induction of oospore formation which generally secretes dose-dependently both α1 and α2 hormones [43,44]. With α1 hormone, sporangia and asexual spores were produced and otherwise are suppressed [43,44]. This mechanism, shared by *Phytophthora* heterothallic species as *P. nicotianae* [45], *P. capsici* [46], *P. cambivora* [47], and *P. infestans* [27], demonstrates that hormones are the universal mating factors among *Phytophthora* species [44]. Unfortunately, according to Tomura et al. [44], certain strains did not produce hormones, and the production time of A1 and A2 mating types were not precisely synchronous [44]. Generally, the *P. infestans* life cycle is suited to its dissemination. Although numerous infections can be caused by asexual sporangia, these are short-lived such as hyphae; instead, oospores can persist for years in soil, surviving freezing and fungicides entailing as an important source of inoculum season by season. In addition, since during biotrophic stages, *P. infestans* can only feed from living cells of the host, often disease diagnosis is delayed. Bypassing plant-based resistance mechanisms, systemic chemicals fungicides remain a key-tool against late blight but are not economically and environmentally sustainable due to the high number of yearly needed applications. In this scenario, aimed at controlling sexual cycle of *P. infestans*, this molecular approach could be a long-term and effective solution to face late blight blocking its sexual reproduction over the years and obtaining positive feedback from farmers and stakeholders [48].

4. Conclusions

The results of this work provide the first evidence for the downregulation of a constitutive gene of *P. infestans* using a recombinant vector based on a plus-sense RNA plant virus suggesting a therapeutic use of the RNAi in the control of threatening pathogens such as *P. infestans*. Furthermore, interference with the sexual reproductive cycle by inhibiting or limiting the production of oospores represents a promising strategy. A partial downregulation of the transcription of *PiLLP* gene involved in oospore formation has been obtained for both tested mating types. The reasons for the partial success need to be addressed. Interestingly, such downregulation was significantly more evident in the A2 mating type as apparently inhibited its growth on PA plates. Although results of this study were not further corroborated by the expected effect on oospore formation, the full exploitation of the approach and future research should primarily focus on the transmissibility of the *PiLLP*-downregulation and fully silenced phenotype to WT isolates during the sexual reproduction. If successful, this approach would benefit of the demonstration by Cao et al. [25] that the nonpersistent acquisition of plant viruses by fungi may commonly occur in nature for its exploitation in the IPM of *P. infestans*.

Supplementary Materials: The following supporting information can be downloaded at: https://www.mdpi.com/article/10.3390/horticulturae9030360/s1, Figure S1: Construction of pTMV-PiLLP-1056 vector to express the 5′UTR + 3′UTR of PiLLP in antisense orientation, under the TMV CP sgRNA promoter.

Author Contributions: Conceptualization, R.L. and T.M.; methodology, R.L. and T.M.; validation, R.L., A.M., R.S. and T.M.; formal analysis, R.L., A.M. and R.S.; investigation, R.L. and T.M.; data curation, R.L.; writing—original draft preparation, R.L. and T.M.; writing—review and editing, R.L., A.M., R.S. and T.M.; All authors have read and agreed to the published version of the manuscript.

Funding: This research received no external funding.

Data Availability Statement: The data presented in this study are available in the article or Supplementary Material.

Acknowledgments: We are grateful to Donato Gallitelli and Antonio Ippolito (University of Bari) for their helpful support in experimental design of research project, critical comments of results, and review of the manuscript. We would like to sincerely thanks Angelo De Stradis (IPSP-CNR, Bari) that kindly provides images within Figure 5.

Conflicts of Interest: The authors declare no conflict of interest. The funders had no role in the design of the study; in the collection, analyses, or interpretation of data; in the writing of the manuscript, or in the decision to publish the results.

References

1. Kamoun, S.; Furzer, O.; Jones, J.D.; Judelson, H.S.; Ali, G.S.; Dalio, R.J.; Roy, S.G.; Schena, L.; Zambounis, A.; Panabières, F. The Top 10 Oomycete Pathogens in Molecular Plant Pathology. *Mol. Plant Pathol.* **2015**, *16*, 413–434.
2. Luan, Y.; Cui, J.; Li, J.; Jiang, N.; Liu, P.; Meng, J. Effective Enhancement of Resistance to *Phytophthora infestans* by Overexpression of MiR172a and b in *Solanum lycopersicum*. *Planta* **2018**, *247*, 127–138. [CrossRef]
3. Fry, W.E.; Birch, P.R.J.; Judelson, H.S.; Grünwald, N.J.; Danies, G.; Everts, K.L.; Gevens, A.J.; Gugino, B.K.; Johnson, D.A.; Johnson, S.B. Five Reasons to Consider *Phytophthora infestans* a Reemerging Pathogen. *Phytopathology* **2015**, *105*, 966–981. [CrossRef]
4. Whisson, S.C.; Vetukuri, R.R.; Avrova, A.O.; Dixelius, C. Can Silencing of Transposons Contribute to Variation in Effector Gene Expression in *Phytophthora infestans*? *Mob. Genet. Elem.* **2012**, *2*, 110–114.
5. Zuluaga, A.P.; Vega-Arreguín, J.C.; Fei, Z.; Matas, A.J.; Patev, S.; Fry, W.E.; Rose, J.K. Analysis of the Tomato Leaf Transcriptome during Successive Hemibiotrophic Stages of a Compatible Interaction with the Oomycete Pathogen *Phytophthora infestans*. *Mol. Plant Pathol.* **2016**, *17*, 42–54. [CrossRef]
6. Whisson, S.C.; Boevink, P.C.; Moleleki, L.; Avrova, A.O.; Morales, J.G.; Gilroy, E.M.; Armstrong, M.R.; Grouffaud, S.; Van West, P.; Chapman, S. A Translocation Signal for Delivery of Oomycete Effector Proteins into Host Plant Cells. *Nature* **2007**, *450*, 115–118.
7. Schornack, S.; van Damme, M.; Bozkurt, T.O.; Cano, L.M.; Smoker, M.; Thines, M.; Gaulin, E.; Kamoun, S.; Huitema, E. Ancient Class of Translocated Oomycete Effectors Targets the Host Nucleus. *Proc. Natl. Acad. Sci. USA* **2010**, *107*, 17421–17426. [CrossRef]
8. Boevink, P.C.; McLellan, H.; Gilroy, E.M.; Naqvi, S.; He, Q.; Yang, L.; Wang, X.; Turnbull, D.; Armstrong, M.R.; Tian, Z. Oomycetes Seek Help from the Plant: *Phytophthora infestans* Effectors Target Host Susceptibility Factors. *Mol. Plant* **2016**, *9*, 636–638. [CrossRef]
9. Lo Presti, L.; Lanver, D.; Schweizer, G.; Tanaka, S.; Liang, L.; Tollot, M.; Zuccaro, A.; Reissmann, S.; Kahmann, R. Fungal Effectors and Plant Susceptibility. *Annu. Rev. Plant Biol.* **2015**, *66*, 513–545.
10. Tyler, B.M.; Tripathy, S.; Zhang, X.; Dehal, P.; Jiang, R.H.; Aerts, A.; Arredondo, F.D.; Baxter, L.; Bensasson, D.; Beynon, J.L. *Phytophthora* Genome Sequences Uncover Evolutionary Origins and Mechanisms of Pathogenesis. *Science* **2006**, *313*, 1261–1266. [CrossRef]
11. Haas, B.J.; Kamoun, S.; Zody, M.C.; Jiang, R.H.; Handsaker, R.E.; Cano, L.M.; Grabherr, M.; Kodira, C.D.; Raffaele, S.; Torto-Alalibo, T. Genome Sequence and Analysis of the Irish Potato Famine Pathogen *Phytophthora infestans*. *Nature* **2009**, *461*, 393–398.
12. Jiang, R.H.; Tyler, B.M. Mechanisms and Evolution of Virulence in Oomycetes. *Annu. Rev. Phytopathol.* **2012**, *50*, 295–318. [CrossRef]
13. Cooke, D.E.; Cano, L.M.; Raffaele, S.; Bain, R.A.; Cooke, L.R.; Etherington, G.J.; Deahl, K.L.; Farrer, R.A.; Gilroy, E.M.; Goss, E.M. Genome Analyses of an Aggressive and Invasive Lineage of the Irish Potato Famine Pathogen. *PLoS Pathog.* **2012**, *8*, e1002940. [CrossRef]
14. Kim, K.-J.; Eom, S.-H.; Lee, S.-P.; Jung, H.-S.; Kamoun, S.; Lee, Y.S. A Genetic Marker Associated with the A1 Mating Type Locus in *Phytophthora infestans*. *J. Microbiol. Biotechnol.* **2005**, *15*, 502–509.
15. Hashemi, M.; Tabet, D.; Sandroni, M.; Benavent-Celma, C.; Seematti, J.; Andersen, C.B.; Grenville-Briggs, L.J. The Hunt for Sustainable Biocontrol of Oomycete Plant Pathogens, a Case Study of *Phytophthora infestans*. *Fungal Biol. Rev.* **2022**, *40*, 53–69.
16. Labarile, R.; Farinola, G.M.; Varsalona, M.; Italiano, F.; Buscemi, G.; Trotta, M.; Grattieri, M. Halotolerance of *Rhodobacter sphaeroides* for Saline and Hypersaline Wastewater Bioremediation. In Proceedings of the 2021 International Workshop on Metrology for the Sea; Learning to Measure Sea Health Parameters (MetroSea), Reggio Calabria, Italy, 4–6 October 2021; pp. 37–42.
17. Grattieri, M.; Labarile, R.; Buscemi, G.; Trotta, M. The Periodic Table of Photosynthetic Purple Non-Sulfur Bacteria: Intact Cell-Metal Ions Interactions. *Photochem. Photobiol. Sci.* **2022**, *21*, 101–111. [PubMed]
18. Buscemi, G.; Labarile, R.; Farinola, G.M.; Trotta, M. Enhancing Photosynthesis in Organic Biological Hybrid Complexes. In *Photochemistry*; Royal Society of Chemistry: Cambridge, UK, 2021; pp. 385–410.
19. Buscemi, G.; Vona, D.; Stufano, P.; Labarile, R.; Cosma, P.; Agostiano, A.; Trotta, M.; Farinola, G.M.; Grattieri, M. Bio-Inspired Redox-Adhesive Polydopamine Matrix for Intact Bacteria Biohybrid Photoanodes. *ACS Appl. Mater. Interfaces* **2022**, *14*, 26631–26641. [PubMed]
20. Su, P.; Tan, X.; Li, C.; Zhang, D.; Cheng, J.; Zhang, S.; Zhou, X.; Yan, Q.; Peng, J.; Zhang, Z.; et al. Photosynthetic bacterium *Rhodopseudomonas palustris* GJ-22 induces systemic resistance against viruses. *Microb Biotechnol.* **2017**, *3*, 612–624. [CrossRef] [PubMed]
21. Zhang, X.; Li, X.; Zhang, Y.; Chen, Y.; Tan, X.; Su, P.; Zhang, D.; Liu, Y. Integrated Control of Potato Late Blight with a Combination of the Photosynthetic Bacterium *Rhodopseudomonas palustris* Strain GJ-22 and Fungicides. *BioControl* **2020**, *65*, 635–645.
22. Whisson, S.C.; Avrova, A.O.; Grenville Briggs, L.J.; van West, P. Mechanisms and Application of Gene Silencing in Oomycetes. In *Oomycete Genetics and Genomics*; John Wiley & Sons, Ltd.: Hoboken, NJ, USA, 2009; pp. 493–515. ISBN 978-0-470-47589-8.
23. Jahan, S.N.; Åsman, A.K.; Corcoran, P.; Fogelqvist, J.; Vetukuri, R.R.; Dixelius, C. Plant-Mediated Gene Silencing Restricts Growth of the Potato Late Blight Pathogen *Phytophthora infestans*. *J. Exp. Bot.* **2015**, *66*, 2785–2794.
24. Duo, J.; Tian, Z.; Liu, J.; Vleeshouwers, V.G.; Shi, X.; Xie, C. Functional Analysis of Potato Genes Involved in Quantitative Resistance to *Phytophthora infestans*. *Mol. Biol. Rep.* **2013**, *40*, 957–967.

25. Cao, X.; Liu, J.; Pang, J.; Kondo, H.; Chi, S.; Zhang, J.; Sun, L.; Andika, I.B. Common but Nonpersistent Acquisitions of Plant Viruses by Plant-Associated Fungi. *Viruses* **2022**, *14*, 2279. [CrossRef] [PubMed]
26. Mascia, T.; Vučurović, A.; Minutillo, S.A.; Nigro, F.; Labarile, R.; Savoia, M.A.; Palukaitis, P.; Gallitelli, D. Infection of *Colletotrichum acutatum* and *Phytophthora infestans* by Taxonomically Different Plant Viruses. *Eur. J. Plant Pathol.* **2019**, *153*, 1001–1017.
27. Mascia, T.; Labarile, R.; Doohan, F.; Gallitelli, D. Tobacco Mosaic Virus Infection Triggers an RNAi-Based Response in *Phytophthora infestans*. *Sci. Rep.* **2019**, *9*, 2657.
28. Mascia, T.; Nigro, F.; Abdallah, A.; Ferrara, M.; De Stradis, A.; Faedda, R.; Palukaitis, P.; Gallitelli, D. Gene Silencing and Gene Expression in Phytopathogenic Fungi Using a Plant Virus Vector. *Proc. Natl. Acad. Sci. USA* **2014**, *111*, 4291–4296. [CrossRef]
29. Feng, H.; Wan, C.; Zhang, Z.; Chen, H.; Li, Z.; Jiang, H.; Yin, M.; Dong, S.; Dou, D.; Wang, Y. Specific Interaction of an RNA-Binding Protein with the 3′-UTR of Its Target MRNA Is Critical to Oomycete Sexual Reproduction. *PLoS Pathog.* **2021**, *17*, e1010001.
30. Situ, J.; Xi, P.; Lin, L.; Huang, W.; Song, Y.; Jiang, Z.; Kong, G. Signal and Regulatory Mechanisms Involved in Spore Development of *Phytophthora* and *Peronophythora*. *Front. Microbiol.* **2022**, *13*, 984672. [PubMed]
31. Guo, T.; Wang, X.-W.; Shan, K.; Sun, W.; Guo, L.-Y. The Loricrin-like Protein (LLP) of *Phytophthora infestans* Is Required for Oospore Formation and Plant Infection. *Front. Plant Sci.* **2017**, *8*, 142.
32. Creager, A.N.; Scholthof, K.-B.G.; Citovsky, V.; Scholthof, H.B. Tobacco Mosaic Virus: Pioneering Research for a Century. *Plant Cell* **1999**, *11*, 301–308. [CrossRef]
33. Shivprasad, S.; Pogue, G.P.; Lewandowski, D.J.; Hidalgo, J.; Donson, J.; Grill, L.K.; Dawson, W.O. Heterologous Sequences Greatly Affect Foreign Gene Expression in Tobacco Mosaic Virus-Based Vectors. *Virology* **1999**, *255*, 312–323. [CrossRef]
34. Wright, K.M.; Cruz, S.S. Life at the Edge-Imaging the Hypersensitive Response Induced by TMV. *Annu. Rep. Scott. Crop Res. Inst.* **2000**, 136–139.
35. Rabindran, S.; Dawson, W.O. Assessment of Recombinants That Arise from the Use of a TMV-Based Transient Expression Vector. *Virology* **2001**, *284*, 182–189. [CrossRef]
36. De Vrieze, M.; Germanier, F.; Vuille, N.; Weisskopf, L. Combining Different Potato-Associated *Pseudomonas* Strains for Improved Biocontrol of *Phytophthora infestans*. *Front. Microbiol.* **2018**, *9*, 2573. [CrossRef]
37. Judelson, H.S. Genetic and Physical Variability at the Mating Type Locus of the Oomycete, *Phytophthora infestans*. *Genetics* **1996**, *144*, 1005–1013. [CrossRef]
38. Kim, K.-J.; Lee, Y.-S. Genetic DNA Marker for A2 Mating Type in *Phytophthora infestans*. *J. Microbiol.* **2002**, *40*, 254–259.
39. Judelson, H.S.; Spielman, L.J.; Shattock, R.C. Genetic Mapping and Non-Mendelian Segregation of Mating Type Loci in the Oomycete, *Phytophthora infestans*. *Genetics* **1995**, *141*, 503–512. [CrossRef] [PubMed]
40. Mazáková, J.; Táborský, V.; Zouhar, M.; Ryšánek, P.; Hausvater, E.; Doležal, P. Occurrence and distribution of mating types A1 and A2 of Phytophthora infestans (Mont.) de Bary in the Czech Republic. *Plant Protect. Sci.* **2006**, *42*, 41–48. [CrossRef]
41. Canto, T.; Palukaitis, P. Novel N Gene-Associated, Temperature-Independent Resistance to the Movement of Tobacco Mosaic Virus Vectors Neutralized by a Cucumber Mosaic Virus RNA1 Transgene. *J. Virol.* **2002**, *76*, 12908–12916. [CrossRef]
42. Lot, H. Contribution a l'etude Du Virus de La Mosaique Du Com-Comber (CMV). I. Method de Purification Rapide Du Virus. *Ann. Phytopathol.* **1972**, *4*, 25–38.
43. Qi, J.; Asano, T.; Jinno, M.; Matsui, K.; Atsumi, K.; Sakagami, Y.; Ojika, M. Characterization of a *Phytophthora* Mating Hormone. *Science* **2005**, *309*, 1828. [CrossRef]
44. Tomura, T.; Molli, S.D.; Murata, R.; Ojika, M. Universality of the *Phytophthora* Mating Hormones and Diversity of Their Production Profile. *Sci. Rep.* **2017**, *7*, 5007. [CrossRef] [PubMed]
45. Cariddi, C.; Mincuzzi, A.; Schena, L.; Ippolito, A.; Sanzani, S.M. First Report of Collar and Root Rot Caused by *Phytophthora nicotianae* on *Lycium barbarum*. *J. Plant Pathol.* **2018**, *100*, 361. [CrossRef]
46. Barchenger, D.W.; Lamour, K.H.; Bosland, P.W. Challenges and Strategies for Breeding Resistance in *Capsicum annuum* to the Multifarious Pathogen, *Phytophthora capsici*. *Front. Plant Sci.* **2018**, *9*, 628. [CrossRef]
47. Vannini, A.; Vettraino, A. *Phytophthora cambivora*. *For. Phytophthoras* **2011**, *1*. [CrossRef]
48. Leesutthiphonchai, W.; Vu, A.L.; Ah-Fong, A.M.; Judelson, H.S. How does *Phytophthora infestans* evade control efforts? Modern insight into the late blight disease. *Phytopathology* **2018**, *108*, 916–924. [CrossRef]

horticulturae

MDPI

Article

Coordinated Regulation of Iron-Acquisition Genes and Citrate Biosynthesis Drives Seasonal Iron Deficiency Adaptation in 'Yali' Pears (*Pyrus bretschneideri* Rehd.)

Shuilin Liu [1,2,†], Ming Zhang [3,†], Huiying Wang [1], Yue Xu [1], Chaodie Wen [3], Jianguang Zhang [1], Yuxing Zhang [1,*] and Haiyan Shi [1,*]

1 College of Horticulture, Hebei Agricultural University, Baoding 071000, China; lsl@bbgu.edu.cn (S.L.); why0108483x@163.com (H.W.); yuexu202209@163.com (Y.X.); zhjg570315@sina.com (J.Z.)
2 College of Resources and Environment, Beibu Gulf University, Qinzhou 535011, China
3 Guangxi Zhuang Autonomous Region Engineering Research Center of Marine Food Nutrition and Processing Technology Innovation, Guangxi College and University Key Laboratory of High-Value Utilization of Seafood and Prepared Food in Beibu Gulf, College of Food Engineering, Beibu Gulf University, Qinzhou 535011, China; zhangming@bbgu.edu.cn (M.Z.); wenchaodie0512@163.com (C.W.)
* Correspondence: yyshhy@hebau.edu.cn or shyrainbow@aliyun.com (H.S.); zhyx@hebau.edu.cn (Y.Z.)
† These authors contributed equally to this work.

Abstract: Iron deficiency chlorosis severely limits the productivity of 'Yali' pears in alkaline soils. This study systematically investigated the physiological and molecular responses of 'Yali' pears to varying degrees of iron deficiency, focusing on the roles of *PbFRO2* (*Pyrus bretschneideri Ferric Reductase Oxidase 2*), *PbIRT1* (*Pyrus bretschneideri Iron-Regulated Transporter 1*), and *PbCS2* (*Pyrus bretschneideri Citrate Synthase 2*) in iron uptake and homeostasis. Based on field observations, pear trees were categorized into normal, moderately chlorotic, and severely chlorotic groups. Results demonstrated that moderate iron deficiency upregulated *PbFRO2* (2.86–7.09-fold), enhanced root ferric reductase (FCR) activity, and promoted Fe^{3+} reduction and Fe^{2+} transport. In contrast, severe deficiency suppressed the expression of these genes and reduced photosynthetic efficiency. Leaf citrate content significantly increased with chlorosis severity, while root citrate content exhibited seasonal fluctuations, peaking in July. Multivariate analyses (PCA and PLS-DA) revealed distinct physiological clustering: normal and moderately chlorotic groups overlapped, whereas the severely chlorotic group formed a separate cluster, reflecting a transition from compensatory activation to metabolic collapse. *PbFRO2* emerged as a central regulator, driving root iron storage in spring and redistribution in summer. These findings elucidate a biphasic adaptation strategy, where moderate deficiency triggers gene-mediated iron mobilization, whereas severe stress disrupts homeostasis. This study provides critical insights into iron metabolism dynamics and proposes *PbFRO2* as a molecular target for breeding iron-efficient pear cultivars.

Keywords: 'Yali' pear; iron deficiency chlorosis; multivariate analysis; seasonal regulation; *PbFRO2*

1. Introduction

Iron is a critical micronutrient essential for plant growth and development, playing a crucial role in the function of many enzymes and proteins. It is involved in important physiological processes such as nitrogen fixation, photosynthesis, respiration, and nucleic acid synthesis. Iron deficiency chlorosis is a global limiting factor for plant growth, manifesting as chlorosis, slow growth, and significant reductions in fruit yield and quality, including soybean (*Glycine max* (L.) Merr.) [1], grapevine (*Vitis vinifera* L.) [2], and citrus (*Citrus* spp.) [3]. In previous reports, the optimal soil pH range for plant iron absorption is typically between 4.0 and 6.0, which falls within the slightly acidic to weakly acidic range. Within this pH range, iron exists primarily in the soluble ferrous form (Fe^{2+}) that can be directly absorbed and utilized by plant roots. When soil pH exceeds 7.0 (neutral or alkaline), iron tends to form insoluble ferric hydroxide (Fe^{3+}) or iron oxide precipitates, significantly reducing its availability and potentially inducing iron-deficiency chlorosis in plants [4,5].

The 'Yali' pear is a traditional and high-quality pear cultivar from North China, prized for its delicate taste and unique flavor [6,7]. However, due to the alkaline and calcareous nature of soils in North China, where iron is predominantly present in an insoluble Fe^{3+} form, the bioavailability of iron is severely limited [8]. As a result, 'Yali' pear trees are particularly susceptible to iron deficiency chlorosis, which significantly impacts their growth, development, fruit quality, and yield.

In plant iron metabolism, the genes *FRO2 (Ferric Reductase Oxidase 2)*, *IRT1 (Iron-Regulated Transporter 1)*, and *CS2 (Citrate Synthase 2)* play crucial roles in the absorption, transport, and storage of iron. First, studies on *Cydonia oblonga* seedlings have revealed that the *FRO2* gene encodes a ferric reductase enzyme that reduces ferric iron (Fe^{3+}) to ferrous iron (Fe^{2+}) in the soil, a process essential for iron absorption by plant roots [9]. Following this, research in *Arabidopsis* has demonstrated that the *IRT1* gene encodes an iron transporter protein responsible for transferring the reduced Fe^{2+} from the soil into root cells, representing a critical step in iron uptake [10,11]. Lastly, investigations on *Malus xiaojinensis* have shown that the *CS2* gene regulates citrate synthesis, which facilitates iron transport and storage. Citrate binds with Fe^{2+} to form soluble complexes, enabling iron translocation across tissues and ensuring its efficient utilization and storage [12]. These genes collectively function through the classical Strategy I mechanism of enhancing Fe^{3+} reduction and solubilization to counteract iron deficiency [13]. The coordinated action of these three genes is essential for the plant's response to iron deficiency, collectively regulating iron metabolism and ensuring that the plant can maintain normal growth and development.

Although agronomic practices such as soil iron application, foliar spraying, and trunk injection can alleviate iron deficiency, these methods typically offer only short-term solutions and fail to fundamentally improve plant iron uptake and utilization efficiency [14]. Therefore, understanding the physiological and molecular mechanisms of plant responses to iron deficiency chlorosis, particularly the processes of iron absorption, transport, and storage, is crucial for addressing iron deficiency issues. This study uses the 'Yali' pear as a model to explore the physiological and molecular responses to varying degrees of iron deficiency chlorosis, with a particular focus on the differential expression of *FRO2*, *IRT1*, and *CS2* genes under iron-deficient conditions. Through a systematic analysis of 'Yali' pear's physiological and molecular responses to different levels of iron deficiency, this research aims to provide a theoretical foundation for studying iron-deficiency mechanisms and for the breeding of iron-deficiency-resistant pear varieties.

2. Materials and Methods

2.1. Plant Materials and Growth Conditions

The field experiment was conducted at the pear orchard of Hebei Agricultural University (38.814° N, 115.422° E). The region has a temperate monsoon climate, with an average temperature of 26 °C and relative humidity of 55% during the experimental period. The soil is classified as calcareous cinnamon soil. The experimental material consisted of seven-year-old 'Yali' pear trees with varying degrees of chlorosis. The trees were planted at a row spacing of 3 m and a tree spacing of 1 m, following conventional fertilizer and water management practices. The experimental trees were categorized into three groups based on the degree of chlorosis: Group 1 (Normal) with no chlorosis, Group 2 (Moderate chlorosis) with leaf interveinal chlorosis and green midribs (25–50% chlorosis of the leaves), and Group 3 (Severe chlorosis) with complete loss of green color, resulting in a yellowish-white appearance (over 75% chlorosis of the leaves). Five replicates were set for each group. To assist in the classification, SPAD (Soil Plant Analysis Development) values were measured at the beginning of the experiment in May. Leaves exhibiting dark green coloration were assigned a SPAD value of 1. Moderately chlorotic leaves had SPAD values approximately 50% of normal, and severely chlorotic leaves had SPAD values below 25% of normal. Given that the normal SPAD value for leaves was 35, moderate chlorotic leaves had SPAD values around 20, and those with severe chlorosis had SPAD values below 10 (Figure 1).

Figure 1. Morphological characteristics of 'Yali' pears under different degrees of iron deficiency.

2.2. SPAD (Soil and Plant Analyzer Development) Measurement

The SPAD meter provides a rapid and non-destructive estimation of relative chlorophyll content, which is used to visually classify the degree of leaf chlorosis. Therefore, in this study, SPAD values were measured using the SPAD-502 PLUS portable chlorophyll meter (Konica Minolta, Osaka, Japan). Measurements were conducted around 10:00 a.m. on clear days. For each tree, six healthy, pest-free leaves from current-year shoots were selected—specifically, the 3rd to 5th fully expanded leaves from the shoot tip—and SPAD measurements were taken at five predetermined positions along each leaf blade. The mean of these five SPAD readings was calculated for each leaf and used as its representative chlorophyll content. The very same leaves used for SPAD determination were employed in all subsequent analyses.

2.3. Chlorophyll Fluorescence Measurement

The leaf chlorophyll fluorescence parameter Fv/Fm was measured using a Handy-PEA fluorimeter Multifunctional Plant Efficiency Analyzer (Hansatech Instruments Ltd., Pentney, Norfolk, UK). Following manufacturer recommendations, leaf samples underwent 20-minute dark adaptation at 25 °C prior to measurement to ensure complete relaxation of all PSII reaction centers. Measurements were performed on the mid-lamina region, an area less susceptible to edge effects and representative of overall photosynthetic performance. Two technical replicates per leaf and three biological replicates per experimental treatment were recorded to ensure data robustness and statistical reliability.

During the Fv/Fm measurement, a 650 nm red LED was used as a saturating pulse light source, with an intensity of 3000 $\mu mol \cdot m^{-2} \cdot s^{-1}$ to determine Fm; F0 was measured under weak measuring light (<1 $\mu mol \cdot m^{-2} \cdot s^{-1}$). No actinic light was used during the Fv/Fm measurements.

2.4. Soil Property Analysis

Soil pH reflects the acidity or alkalinity of the soil solution, which is a key factor in nutrient uptake and root activity. In this study, rhizosphere soil attached to roots during sampling was used for pH analysis [15]. Air-dried soil (passed through a 2 mm nylon sieve) was weighed (10 g) into a 100 mL Erlenmeyer flask, mixed with 25 mL distilled water, stirred vigorously with a glass rod for 2 min, and allowed to settle for 30 min. pH values were measured using a pH meter (PB-10, Sartorius Scientific Instruments Co., Ltd., Beijing, China), with three replicates per treatment.

Soil organic matter is essential for enhancing soil fertility, structure, and water retention, thereby supporting nutrient cycling and overall ecosystem sustainability. In this study, the potassium dichromate-concentrated sulfuric acid method was used for soil organic matter analysis [16]. Briefly, 0.5 g of rhizosphere soil (passed through a 100-mesh nylon sieve) was placed in a 250 mL Erlenmeyer flask. Then, 10 mL of 1 mol/L $K_2Cr_2O_7$ (1/6 equivalent concentration, (Shanghai Macklin Reagent Co. Ltd., Shanghai, China, all reagents used herein are designated as Macklin Reagent unless otherwise specified).) and 20 mL concentrated H_2SO_4 (Sinopharm Chemical Reagent Co. Ltd., Shanghai, China) were added. The mixture was slowly rotated for 1 min, incubated for 30 min, diluted with 220 mL ultrapure water, and titrated with 0.5 mol/L $FeSO_4$ after adding 3–4 drops of phenanthroline indicator until the solution turned brick-red. Calculations were performed according to standard formulae.

Available iron refers to the proportion of iron that can be absorbed by plants. In this study, the DTPA extraction method was used for available iron analysis [17]. Specifically, 10.0 g of air-dried rhizosphere soil (2 mm nylon-sieved) was mixed with 20.0 mL of DTPA extractant (pH 7.30; containing 0.005 mol/L DTPA, 0.01 mol/L $CaCl_2$, and 0.1 mol/L triethanolamine, adjusted to pH 7.30 with 6 mol/L HCl) in a 100 mL Erlenmeyer flask. The suspension was shaken at 180 rpm for 2 h at 25 °C, immediately filtered, and the filtrate was analyzed by ICP-AES (PerkinElmer Optima 8000, Waltham, MA, USA). A standard calibration curve was established using iron reference materials prior to quantification.

2.5. Root Vitality Measurement

Root vitality was determined as an indicator of metabolic activity in root tissues, reflecting the capacity for nutrient uptake and stress adaptation [18]. The assay was performed using the triphenyltetrazolium chloride (TTC) reduction method [19]. Root sampling was performed at four cardinal directions (8 times trunk diameter distance from the tree base) using steel shovels. After removing surface soil, roots from a 20–40 cm depth were carefully excavated through soil sieving. Secondary lateral roots (diameter <5 mm) with adhering rhizosphere soil were collected for physiological analysis. Approximately 0.5 g of root tissue from pear trees was incubated with 0.4% TTC solution and 5 mL sulfuric acid buffer (pH = 7), and the root tips were fully immersed. The samples were incubated at 37 °C in the dark for 2 h, and the reaction was then terminated by adding 2 mL of 1 mol/L H_2SO_4. The roots were ground with ethyl acetate, washed 3 times, and triphenylformazan (TTF) was extracted. The absorbance at 485 nm was measured using a spectrophotometer (UV-2550, Shimadzu, Kyoto, Japan). Each sample was repeated 3 times.

2.6. Ferric-Chelate Reductase (FCR) Enzyme Activity Measurement

FCR activity was measured according to Zheng et al. [20], utilizing the same root specimens collected for root vitality assessment. Root segments were incubated in a $CaSO_4$ (0.5 mM) solution for 10 min. After washing and wiping, they were placed in 5 mL of enzyme activity detection solution containing 50 mM Tris (pH 7.5), 0.5 mM $CaSO_4$, 0.3 mM BPDS, and 50 mM FeEDTA. The reaction was carried out in the dark at 25 °C for 1 h, with gentle inversion every 10 min to mix. After the reaction, 2 mL of the colored supernatant was transferred to a 1 cm quartz cuvette and measured at 535 nm using a UV-2550 UV-Vis spectrophotometer (Shimadzu, Kyoto, Japan). A standard curve was established before measuring the samples. Each sample was repeated 3 times.

2.7. Total and Active Iron Measurement in Leaves

Total iron and active iron content were measured using an inductively coupled plasma atomic emission spectrometer (ICP-AES, PerkinElmer Optima 8000, Waltham, MA, USA), with slight modifications to the method of Lin et al. [21]. Prior to sample analysis, calibration procedures were carried out; wavelength calibration was performed using the instrument's built-in software system, and analytical calibration was achieved by constructing a standard calibration curve, ensuring that measurements were conducted within the curve's linear range. Leaf samples were dried at 105 °C for 40 min and further dried at 75 °C to constant weight. After grinding, 1 g of dried leaf powder was digested with 20 mL HNO_3 (Sinopharm Chemical Reagent Co. Ltd., Shanghai, China) and 5 mL $HClO_4$ (Sinopharm Chemical Reagent Co. Ltd., Shanghai, China), and the solution was diluted to 50 mL. The total iron content was then quantified using an ICP-AES.

Active iron content in fresh leaf or root samples (0.25 g) was extracted with 10 mL 1 mol/L HCl on a shaker for 12 h, filtered, diluted to 25 mL, and measured by ICP-AES.

2.8. Measurement of Citrate in New Roots and Leaves

Citrate content was quantified by high-performance liquid chromatography (HPLC) following the protocol of Lucarini et al. [22]. Briefly, approximately 0.2 g of fresh roots or leaves were homogenized in liquid nitrogen, followed by extraction with ddH_2O to a final volume of 1.5 mL. The mixture was incubated at 75 °C for 30 min and subsequently centrifuged at 12,000× *g* for 30 min. After centrifugation, the supernatant was filtered through a 0.22 μm membrane prior to HPLC analysis.

Chromatographic conditions were as follows: An Agilent 1260 HPLC system (Agilent Technologies, Santa Clara, CA, USA) equipped with a UV detector was employed. Separations were achieved using an isocratic method with an Eclipse Plus C18 column (4.6 mm × 250 mm, 5 μm particle size) maintained at 25 °C. The mobile phase consisted of 0.02 mol/L potassium dihydrogen phosphate buffer (pH 2.8), which was delivered isocratically at a flow rate of 0.5 mL/min. Finally, a 10 μL aliquot of the processed sample was injected for each analysis.

2.9. qRT-PCR Analysis

Total RNA Extraction: Total RNA was extracted from pear roots using the Tiangen RNA extraction kit (TIANGEN, Beijing, China) following the manufacturer's instructions. RNA purity (A260/A280 ratio 1.8–2.0) was confirmed by UV spectrophotometry, and integrity was verified by electrophoresis on 1.2% agarose gels.

Genomic DNA removal and cDNA synthesis: Genomic DNA removal and first-strand cDNA synthesis were performed strictly following the manufacturer's instructions of the Fast QuantRT Kit (with gDNase) (TIANGEN, Beijing, China). The detailed procedures were as follows: a 10 μL reaction mixture was prepared containing 2.0 μL of 5×gDNA Buffer, 1.0 μg of total RNA, and RNase-free ddH_2O to adjust the volume. After brief centrifugation, the mixture was incubated at 42 °C for 3 min and immediately placed on ice. For reverse transcription, 10 μL of RT master mix containing 2.0 μL of 10 × Fast RT Buffer, 1.0 μL of RT Enzyme Mix, 2.0 μL of FQ-RT Primer Mix, and RNase-free ddH_2O to a final volume of 20 μL was added to the gDNA-removed reaction solution. After thorough mixing, the reaction was performed at 42 °C for 15 min, followed by heat inactivation at 95 °C for 3 min.

qRT-PCR analysis: qRT-PCR was performed using the Mastercycler ep realplex4 system (Eppendorf, Hamburg, Germany) with TransStart Top Green qPCR SuperMix (AQ131, TransGen Biotech, Beijing, China). The 20 μL PCR reaction mixture contained 10 μL of TransStart SuperMix, 2.0 μL of cDNA template, 0.4 μL each of upstream and downstream primers (10 μM), and 7.2 μL of ddH_2O. The thermal cycling conditions were: initial denaturation at 94 °C for 30 s; 42 cycles of denaturation at 94 °C for 5 s, annealing at 55 °C for 15 s, and extension at 72 °C for 10 s; followed by melting curve analysis (95 °C for 15 s, 55 °C for 60 s, and 95 °C for 15 s). All samples were analyzed with three technical replicates. The pear actin gene was used as an internal reference, and relative gene expression levels were calculated using the $2^{-\Delta\Delta CT}$ method.

Primer Design: Specific primers for pear genes related to iron absorption and transport were designed by querying the NCBI database. The primers were synthesized by GeneScript Biotech (Shanghai) Co., Ltd. (Shanghai, China). The gene names and sequences are listed in Table 1.

Table 1. Primer sequences used for qPCR analysis.

Gene	Primer Sequences	
PbACT	qPbAcT-F	5′-TTGGTATGGGTCAGAAGG-3′
	qPbAcT-R	5′-CTGTGAGCAGAACTGGGTG-3′
PbFRO2	q*PbFRO2*-F	5′-GCGGAGGAAGCGGTATTA-3′
	q*PbFRO2*-R	5′-CCGAAACGGAAGGATTA-3′
PbIRT1	q*PbIRT1*-F	5′-AGTCTGAGGAGGACGAGTGC-3′
	q*PbIRT1*-R	5′-GGAACGAGTGACAAGAGGGA-3′
PbCS2	q*PbCS2*-F	5′-GGCGAGAGAAAGAGAGGTGATC-3′
	q*PbCS2*-R	5′-CGGTGACGCTCCTGAAGAATAC-3′

2.10. Statistical Analysis

All experimental data were organized using Microsoft Excel (Microsoft 365, Redmond, WA, USA) to ensure accuracy and completeness, with each treatment group being repeated three times, and the results are presented as mean ± standard deviation. Statistical significance between treatment groups was analyzed using SPSS (IBM, SPSS Statistics, version 25), with one-way analysis of variance (ANOVA) for group comparisons. When significant differences were found, Tukey's HSD test was applied for post-hoc multiple comparisons, with a significance level set at $p < 0.05$. Data visualization was performed using GraphPad Prism (GraphPad Software, version 12.5), including bar charts, line graphs, etc., with significance markers added based on SPSS results. Additionally, correlation analysis, principal component analysis (PCA), and Variable Importance in Projection for Partial Least Squares Discriminant Analysis (PLS-DA VIP) were conducted using the MetaboAnalyst website (https://www.metaboanalyst.ca/ (accessed on 2 February 2025)), revealing the relationships between variables and the distribution characteristics of the samples across different treatments.

3. Results

3.1. Physicochemical Properties of Soil

To investigate the relationship between soil physicochemical properties and the degree of chlorosis in 'Yali' pear, we analyzed the pH, available iron content, and organic matter content of the rhizosphere soil in different treatment groups. The results (see Table 2) showed that from May to July, the pH of the rhizosphere soil of moderate chlorosis and severe chlorosis pear trees was significantly higher than that of normal trees. In May, the soil pH was at its lowest, with values of 7.41, 7.76, and 7.86 for the normal, moderate, and severe chlorosis groups, respectively; by July, the pH values peaked at 7.78 for the normal group, 7.92 for the moderate group, and 8.0 for the severe group. In addition, the available iron content in the rhizosphere of chlorotic pear trees was significantly lower than that in the normal group, particularly in the severe chlorotic group, where the minimum value was only 6.94 mg/kg, compared to the normal group's lowest available iron content of 20.11 mg/kg in June. The organic matter content in May was significantly lower in the moderate and severe chlorosis groups than in the normal group; no significant differences were observed in June; however, in July, the severe chlorosis group was significantly lower than the normal group. In summary, even though the overall soil pH in this study fell outside the optimal range (pH 4–6) for iron availability and solubility, our findings still demonstrated that increased iron-deficiency chlorosis severity correlated with elevated rhizosphere soil pH and reduced available iron content. This suggests these parameters may constitute key drivers of iron deficiency-induced chlorosis in pear trees.

Table 2. Rhizosphere soil physical and chemical properties of pear trees under different iron deficiency conditions.

Sample Dates	Chlorisis Types	Soil pH Value	Soil Active Iron ($mg \cdot kg^{-1}$)	Organic Matter Content ($mg \cdot kg^{-1}$)
5 May	Normal	7.41 ± 0.04 c	35.2 ± 1.52 a	26.37 ± 1.30 a
	Moderate chlorosis	7.76 ± 0.05 b	29.05 ± 1.39 b	18.34 ± 0.86 b
	Severe chlorosis	7.86 ± 0.05 a	7.99 ± 0.15 c	19.95 ± 0.56 b
5 June	Normal	7.63 ± 0.01 b	20.11 ± 1.17 a	21.10 ± 1.62 a
	Moderate chlorosis	7.91 ± 0.02 a	7.88 ± 0.10 b	21.09 ± 2.13 a
	Severe chlorosis	8.00 ± 0.05 a	6.94 ± 0.16 b	18.11 ± 0.86 a
5 July	Normal	7.78 ± 0.06 b	52.85 ± 0.50 a	26.60 ± 0.65 a
	Moderate chlorosis	7.92 ± 0.07 a	22.29 ± 0.82 b	23.62 ± 2.53 ab
	Severe chlorosis	8.04 ± 0.02 a	14.02 ± 0.78 c	20.87 ± 0.32 b

Note: Different letters indicate significant differences at the $p < 0.05$ level.

3.2. Correlation Between Different Chlorosis Levels and Iron Content

To further explore the relationship between iron deficiency and leaf chlorosis, we measured both the active and total iron content in leaves and roots. The results showed that the total iron content in the leaves did not exhibit a consistent trend over the experimental period (see Figure 2A), whereas active iron content decreased significantly with increasing chlorosis severity (see Figure 2B). In all months, the active iron content in the leaves of the normal group was significantly higher than that of the moderate and severe chlorosis groups. Furthermore, as shown in Figure 2C, the ratio of active iron to total iron in the leaves followed the same trend as the changes in active iron content, consistent with previous studies, which suggest that the ratio of active iron to total iron can reflect the overall iron utilization level of the plant [23]. These results further indicate that active iron content is an important indicator for evaluating the degree of iron deficiency-induced chlorosis. As shown in Figure 2D, the active iron content in the roots generally showed a trend of first increasing and then decreasing, peaking in June. From May to July, the active iron content in the roots of the normal and moderate groups was significantly higher than that of the severe chlorosis group. In May and June, there was no significant difference in the root active iron content between the normal and moderate chlorosis groups, but in July, the normal group had significantly higher active iron content than the moderate chlorosis group. Overall, the total iron content in the leaves did not show a clear correlation with the degree of chlorosis, but the measurement of active iron content in both the leaves and roots showed a decreasing trend as leaf chlorosis intensified, further suggesting that active iron content is an important indicator for assessing the degree of iron chlorosis.

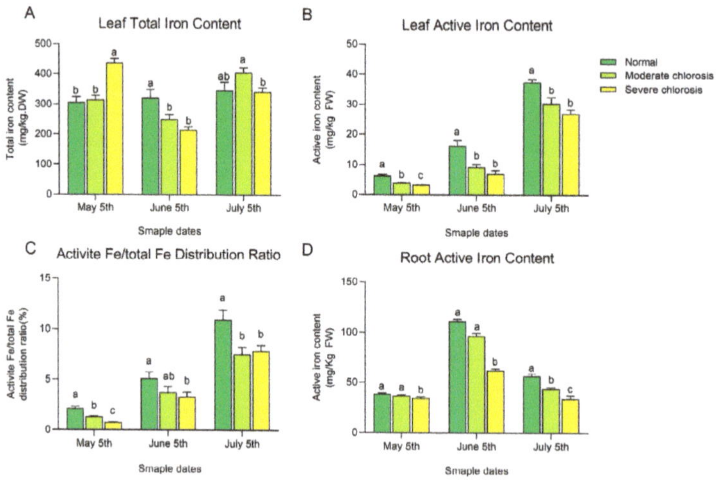

Figure 2. Iron content in roots and leaves of pear trees under different iron deficiency conditions. (**A**) Total iron content in leaves. (**B**) Active iron content in leaves. (**C**) Active iron content in leaves as a percentage of total iron. (**D**) Active iron content in roots. Different letters indicate significant differences at the $p < 0.05$ level.

3.3. Correlation Between Different Chlorosis Levels and Relative Chlorophyll Content and Chlorophyll Fluorescence Parameters

SPAD values, which serve as a relative indicator of chlorophyll content, were used to assess leaf chlorosis [24]. The results (see Figure 3A) indicated that SPAD values decreased significantly with the aggravation of chlorosis. The SPAD values of the normal group were 1.87 and 10.26 times, 1.70 and 8.88 times, and 1.57 and 3.59 times those of the moderate and severe chlorosis groups in each respective month. The maximum photochemical efficiency (Fv/Fm), reflecting the light energy conversion efficiency of PSII reaction centers, showed minor variations under normal conditions but decreased significantly under iron stress [25]. Results demonstrated that Fv/Fm values remained relatively stable in the normal group, with no significant differences observed between moderately chlorotic and normal leaves across months. However, Fv/Fm values in severely chlorotic leaves were significantly lower than those in both normal and moderately chlorotic groups. The Fv/Fm values of severely chlorotic leaves dropped to 0.16 in May and remained around 0.6 from June to July. These findings indicate that both moderate and severe iron deficiency chlorosis can reduce relative chlorophyll content, but only severe chlorosis significantly inhibits photosynthesis. The concurrent decline in SPAD values and Fv/Fm ratio under severe iron deficiency may be attributed to impaired iron mobility within plants, particularly given that our sampling focused on the 3rd to 5th newly emerged leaves at shoot apices—positions subject to lower iron allocation priority [26]. In contrast, the observed recovery of Fv/Fm values in June and July likely correlates with the dynamic accumulation of active iron content in leaves during developmental progression (Figure 3B).

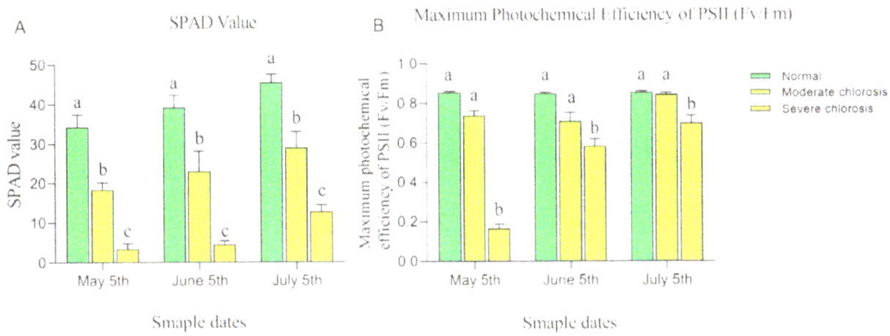

Figure 3. Chlorophyll content and maximum photochemical efficiency (Fv/Fm) of young leaves in pear trees under different iron deficiency conditions. (**A**) SPAD value. (**B**) Maximum photochemical efficiency of PSII (Fv/Fm). Different letters indicate significant differences at the $p < 0.05$ level.

3.4. Correlation Between Different Chlorosis Levels and Root Iron Uptake Ability

Previous studies have shown that root vitality is an important indicator for evaluating a plant's capacity to absorb nutrients [27]. In this study, the root vitality of 'Yali' pears under different chlorosis levels was measured (see Figure 4A).

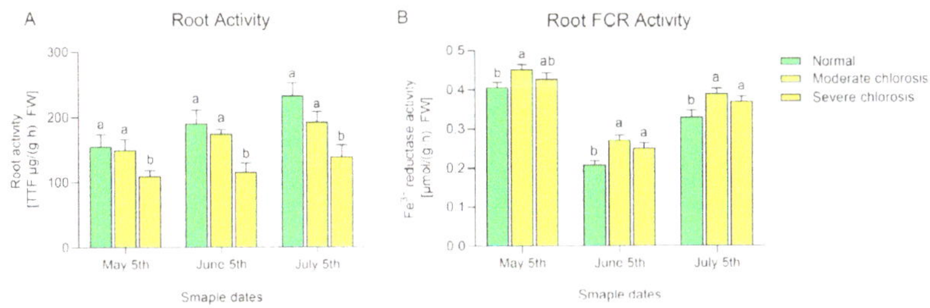

Figure 4. Root activity and ferric reductase activity of pear trees under different iron deficiency conditions. (**A**) root activity. (**B**) Root FCR activity. Different letters indicate significant differences at the $p < 0.05$ level.

The results indicated that during the experimental period, there was no significant difference in root vitality between the normal and moderate groups, but both were significantly higher than the severe chlorosis group. Pear trees need to reduce Fe^{3+} to Fe^{2+} in their roots for effective absorption, so the activity of root FCR can reflect the plant's iron absorption ability [28]. As shown in Figure 4B, from May to July, the activity of root FCR in all groups exhibited a trend of first decreasing and then increasing.

In each month, the FCR activity of the moderate chlorosis group was significantly higher than that of the normal group, and slightly higher than that of the severe chlorosis group, though there was no significant difference between the two; in May, there was no significant difference between the normal and severe chlorosis groups, but later, the normal group became significantly lower than the severe chlorosis group. Overall, the intensification of chlorosis led to a significant decrease in root vitality; within the experimental period, FCR activity showed a trend of first increasing and then decreasing as the degree of chlorosis deepened.

3.5. Effect of Different Chlorosis Levels and Citrate Content

Citrate, as an organic acid, has been shown to play an important role in plant tolerance to abiotic stress, and the long-distance transport of iron within plants is also related to citrate content [29]. In this study, the citrate content in both leaves and roots was measured. As shown in Figure 5A, within the experimental range, the citrate content in the leaves significantly increased with the intensification of chlorosis. In all months, the severe chlorosis group had significantly higher citrate content than the moderate chlorosis group, and the moderate chlorosis group had significantly higher content than the normal group. The citrate content in the severe chlorosis group was 4.48 times, 1.81 times, and 4.16 times that of the normal group in May, June, and July, respectively. This suggests that low active iron content in the leaves induces an increase in citrate content. In the analysis of citrate content in the roots (Figure 5B), it was found that from May to June, the citrate content in the roots of the normal plants was significantly lower than that of the moderate and severe chlorosis groups. In July, the normal group had the highest citrate content in the roots, followed by the moderate chlorosis group, and the severe chlorosis group had the lowest. These results suggest that the variation in citrate content in the roots may be influenced by seasonal factors, and plants may adopt different strategies to regulate citrate synthesis and accumulation during different growth stages.

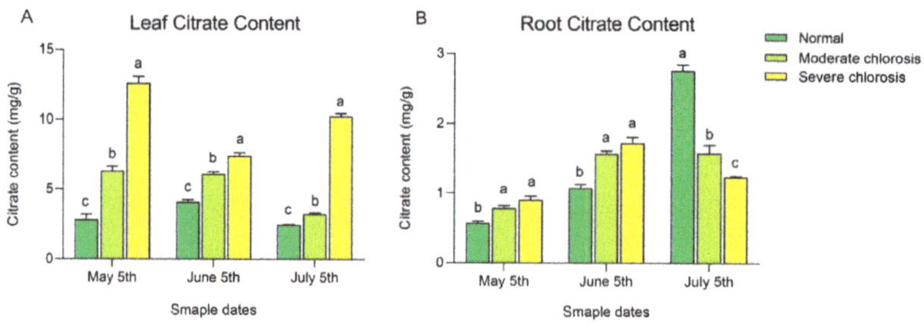

Figure 5. Citrate content in roots and leaves of pear trees under different iron deficiency conditions. (**A**) Citrate content in leaves. (**B**) Citrate content in roots. Different letters indicate significant differences at the $p < 0.05$ level.

3.6. Correlation Between Different Chlorosis Levels and the Expression of Iron Uptake-Related Genes

To further investigate the impact of different chlorosis levels on the expression of genes involved in iron uptake and transport, we examined the expression levels of *PbCS2 (Pyrus bretschneideri Citrate Synthase 2)*, *PbFRO2 (Pyrus bretschneideri Ferric Reductase Oxidase 2)*, and *PbIRT1 (Pyrus bretschneideri Iron-Regulated Transporter 1)* in the roots of 'Yali' pears (see Figure 6). In this study, root expression of *PbCS2* gradually decreased from May to July, while its expression in the moderate chlorosis group was significantly higher than in both the normal and severe chlorosis groups, with fold changes of 8.09 and 3.46, 2.88 and 2.25, and 2.05 and 2.50, respectively. As shown in Figure 6B, from May to July, root *PbFRO2* expression in the moderate chlorosis group was significantly higher than in both the normal and severe groups, with fold differences of 2.86 and 7.09, 3.62 and 6.99, and 3.49 and 5.97, respectively; the normal group exhibited slightly higher *PbFRO2* expression than the severe group, though the difference was not significant. This indicates that moderate chlorosis may be the critical threshold for inducing high *PbFRO2* expression, which helps enhance iron uptake in the roots. As shown in Figure 6C, in May and June, the expression of *PbIRT1* in the moderate and severe chlorosis groups was significantly higher than in the normal group, with fold differences of 20.25 and 9.75, and 4.88 and 2.07, respectively, and the moderate group was significantly higher than the severe group; in July, the expression in the moderate group was significantly higher than in both the normal and severe groups. Our findings revealed a significant negative correlation between iron deficiency status and root iron translocation capacity in pear trees, with enhanced root iron transport activity under moderate deficiency but markedly reduced capacity observed during severe leaf chlorosis.

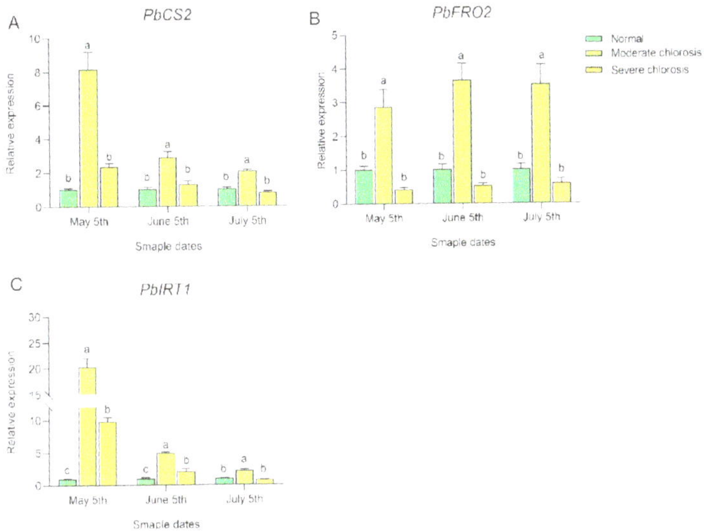

Figure 6. Expression levels of iron uptake-related genes in roots of pear trees under different iron deficiency conditions. (**A**) Expression level of *PbCS2*. (**B**) Expression level of *PbFRO2*. (**C**) Expression level of *PbIRT1*. Different letters indicate significant differences at the $p < 0.05$ level.

3.7. Multivariate Analysis of Physiological Traits and Gene Expression

Principal Component Analysis (PCA) is an unsupervised statistical method that transforms high-dimensional data into a smaller number of principal components, thereby revealing overall differences and distribution patterns among samples. Partial Least Squares Discriminant Analysis (PLS-DA) is a supervised multivariate analysis technique used to identify variables that effectively distinguish between different treatment groups, while the Variable Importance in Projection (VIP) score is used to evaluate the importance of each variable in the classification model.

Therefore, to further elucidate the physiological changes and molecular regulatory mechanisms of pear trees under iron deficiency stress, we performed multivariate analysis on the measured physiological parameters and the expression levels of iron metabolism-related genes. We applied PCA and PLS-DA analyses, combined with VIP scoring, to assess the contribution of each variable to group discrimination. Through these methods, we aimed to uncover the effects of chlorosis severity on iron uptake, transport, and utilization in pear trees, with particular emphasis on the roles of the *PbCS2*, *PbIRT1*, and *PbFRO2* genes.

3.7.1. PCA of Physiological Traits and Gene Expression

As shown in Figure 7A, when PCA was performed based solely on chlorosis severity without distinguishing sampling periods, normal and moderately chlorotic leaves were not well separated, whereas severely chlorotic leaves exhibited clear differentiation from the other two groups. This result suggests that normal and moderately chlorotic leaves share similar physiological and molecular characteristics, while severe chlorosis induces more pronounced physiological alterations.

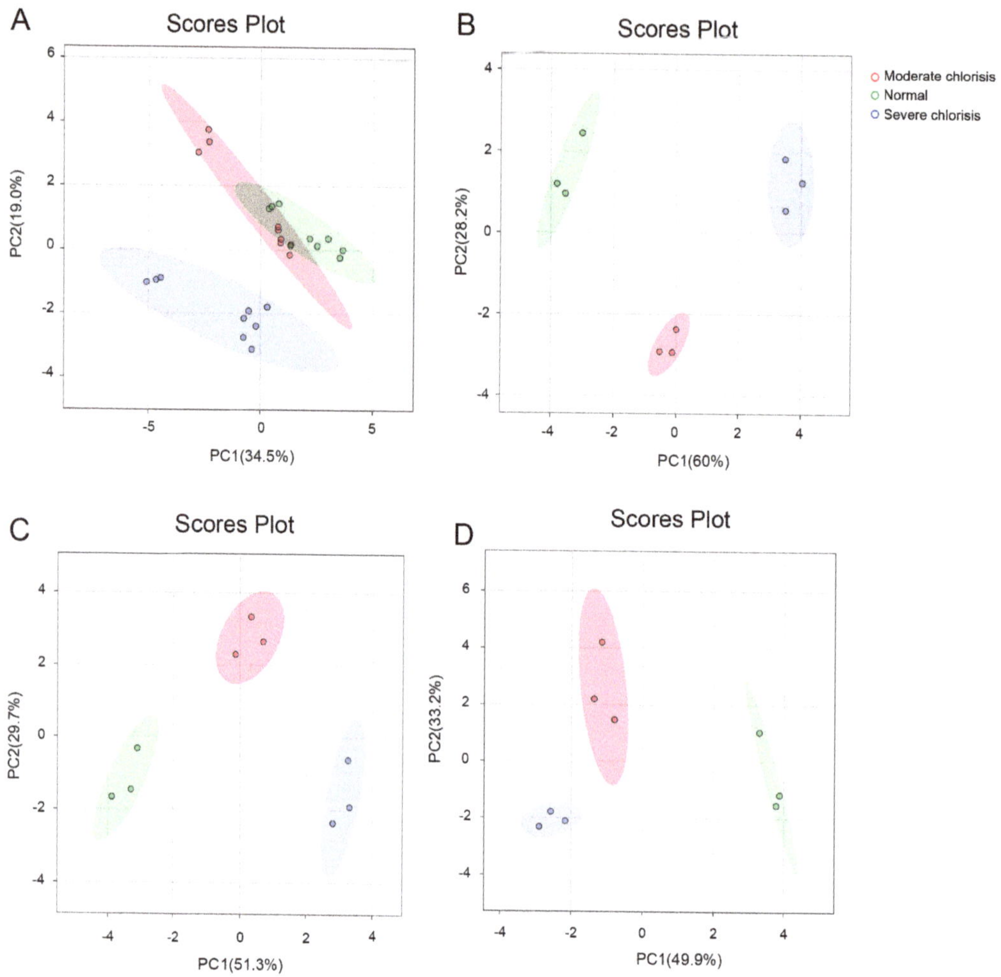

Figure 7. PCA of 'Yali' pears at different sampling times under different degrees of iron deficiency. (**A**) All samples grouped by chlorosis severity, regardless of sampling time. (**B**) Data from May. (**C**) Data from June. (**D**) Data from July.

Figure 7B–D shows the results of PCA analysis for different chlorosis groups from May to July. The results indicated that the sample distribution patterns for all three sampling months were consistent: along the PC1 axis, the moderate chlorosis samples were always positioned between the normal and severe chlorosis samples, with clear separation among the three groups. This spatial arrangement along the PC1 axis suggests that the physiological transition from normal to severe chlorosis follows a gradient trajectory rather than a phenotypic shift, reflecting the cumulative metabolic disturbances caused by severe iron deficiency. Specifically, as shown in Table 3, the loading values for SPAD, which reflects leaf color and photosynthetic capacity, the citrate content in both the roots and leaves, which reflects iron transport capacity, and the available iron in the soil (soil active iron content) were all greater than 0.25 in the different months.

Table 3. Identified physiological and molecular indicators with high absolute loadings on PC1 and PC2.

May to July				May			
Physiological /Molecular Indicators	**PC1 Loadings**	**Physiological /Molecular Indicators**	**PC2 Loadings**	**Physiological /Molecular Indicators**	**PC1 Loadings**	**Physiological /Molecular Indicators**	**PC2 Loadings**
Total iron-leaf	-0.354	*PbCS2*	0.371	SPAD	−0.320	*PbCS2*	−0.452
SPAD	0.329	Citrate-root	−0.333	Citrate-leaf	0.318	*PbIRT1*	−0.424
Citrate -leaf	−0.325	*PbIRT1*	0.329	Active iron/Total iron	−0.312	*PbFRO2*	−0.420
Root activity	0.325	Available iron-soil	0.323	Available iron-soil	−0.308	FCR activity	−0.372
FCR activity	−0.317	Citrate -leaf	−0.319	Fv/Fm	−0.305	Soil pH	−0.315
Active iron/Total iron	0.294	*PbFRO2*	0.309	Active iron-leaf	−0.304	Soil organic matter	0.270
PbIRT1	−0.282	SPAD	0.294	Active iron-root	−0.302	Total iron-leaf	0.171
Active iron-leaf	0.269	Soil pH	−0.279	Total iron-leaf	0.292	Citrate-root	−0.159
Fv/Fm	0.255	Active iron/Total iron	−0.244	Citrate-root	0.292	Fv/Fm	−0.136
PbCS2	−0.208	Active iron-leaf	−0.175	Root activity	−0.268	Root activity	−0.132
Available iron-soil	0.176	Soil organic matter	−0.147	Soil organic matter	−0.247	Active iron-leaf	0.119
Citrate-root	0.164	Active iron-root	−0.139	*PbIRT1*	0.120	Available iron-soil	−0.111
Soil pH	−0.158	FCR activity	0.116	*PbFRO2*	−0.089	Active iron-root	−0.097
Active iron-root	0.147	Fv/Fm	0.111	*PbCS2*	0.032	Active iron/Total iron	0.044
Soil organic matter	0.045	Root activity	0.105	FCR activity	−0.020	SPAD	0.029
PbFRO2	0.031	Total iron-leaf	−0.095	Soil pH	0.014	Citrate-leaf	0.022

June				July			
Physiological /Molecular Indicators	**PC1 Loadings**	**Physiological /Molecular Indicators**	**PC2 Loadings**	**Physiological /Molecular Indicators**	**PC1 Loadings**	**Physiological /Molecular Indicators**	**PC2 Loadings**
FCR activity	0.341	*PbFRO2*	0.439	Available iron-soil	0.350	Active iron/Total iron	−0.377
Citrate-leaf	0.341	*PbCS2*	0.371	Total iron-leaf	−0.345	*PbFRO2*	0.375
Citrate-root	0.339	*PbIRT1*	0.350	SPAD	0.342	*PbIRT1*	0.367
Soil pH	0.334	Total iron-leaf	−0.331	Citrate-root	0.333	*PbCS2*	0.363
SPAD	−0.333	Active iron-root	0.327	FCR activity	−0.325	Active iron-leaf	−0.332
Soil organic matter	0.302	Root activity	0.291	Active iron-root	0.286	Soil organic matter	−0.330
Available iron-soil	−0.286	Available iron-soil	−0.256	Root activity	0.283	Citrate-leaf	−0.275
Fv/Fm	0.280	Active iron/Total iron	−0.251	Soil pH	−0.277	Soil pH	−0.265
Active iron-leaf	−0.257	Active iron-leaf	−0.229	Citrate-leaf	−0.259	Fv/Fm	−0.195
PbIRT1	0.191	Fv/Fm	−0.155	Fv/Fm	−0.211	FCR activity	−0.137
PbCS2	0.172	Soil pH	−0.120	Active iron-leaf	0.150	Citrate-root	−0.092
Active iron-root	−0.170	Soil organic matter	0.012	Active iron/Total iron	0.132	Root activity	0.094
Total iron-leaf	0.128	SPAD	0.117	*PbCS2*	−0.127	SPAD	0.077
Root activity	0.022	Citrate-leaf	−0.094	*PbIRT1*	−0.104	Active iron-root	−0.059
Active iron/Total iron	0.020	Citrate-root	−0.067	*PbFRO2*	−0.082	Available iron-soil	−0.043
PbFRO2	0.016	FCR activity	0.002	Soil organic matter	−0.005	Total iron-leaf	0.022

In contrast, along the PC2 axis, the normal and severe chlorosis samples were almost indistinguishable, while the moderate chlorosis samples were clearly separated. This pattern suggests that PC2 captures the nonlinear physiological responses during chlorosis, characterized by an initial increase followed by a decrease. According to Table 3, PC2 was primarily influenced by the activity of iron metabolism-related enzymes, including the expression levels of *PbCS2*, *PbIRT1*, and *PbFRO2* (Loading value > 0.35). During the early stages of chlorosis, these genes are upregulated to enhance iron acquisition and redistribution. However, as iron deficiency continues, these regulatory mechanisms gradually weaken or become suppressed, potentially exacerbating the severity of chlorosis. These genes are upregulated in the early stages of chlorosis to enhance iron acquisition and redistribution. However, as iron deficiency persists, these regulatory mechanisms gradually weaken or become suppressed.

Overall, these findings demonstrate a dual-phase physiological adaptation strategy in pear trees under iron-deficient conditions: PC1 represents the primary and progressive physiological alterations, whereas PC2 reflects a transient adaptive mechanism that is initially activated and later declines. This distinction underscores the complex metabolic regulatory strategies employed by pear trees under iron deficiency stress.

3.7.2. VIP Analysis of Key Variables Contributing to Chlorosis and Iron Deficiency Responses

The PLS-DA VIP analysis identified *PbFRO2* as a consistently significant variable (VIP > 1.5) associated with leaf chlorosis severity across all sampled growth stages (Figure 8). In the overall analysis without distinguishing sampling months (Figure 8A), *PbFRO2* exhibited the highest VIP value (2.1331), indicating its significant contribution to iron deficiency-induced chlorosis across all time points. In the month-specific analysis (Figure 8B–D), *PbFRO2* had the highest VIP values in May and July (1.544 and 1.531, respectively), while it ranked second in June (1.584), suggesting that *PbFRO2* functions as a central regulatory factor in iron deficiency responses, particularly during early and late growth stages. Moreover, in May, leaf total iron content, Fv/Fm, and *PbCS2* had high contributions, whereas in June, *PbIRT1*, leaf Citrate, and *PbCS2* became more influential. By July, root active iron, *PbFRO2*, and soil pH had the highest contributions (Table 4). This dynamic shift suggests that the iron deficiency response in pear trees is not static but undergoes seasonal regulation to adapt to changing environmental conditions.

Table 4. List of response variables with VIP values greater than 1.

May to July		May		June		July	
Physiological /Molecular Indicators	**VIP Value**	**Physiological /Molecular Indicators**	**VIP Value**	**Physiological /Molecular Indicators**	**VIP Value**	**Physiological /Molecular Indicators**	**VIP Value**
PbFRO2	2.133	*PbFRO2*	1.544	Active iron-root	1.617	*PbFRO2*	1.584
Citrate-leaf	1.592	Total iron content-leaf	1.347	*PbFRO2*	1.531	*PbIRT1*	1.564
SPAD	1.258	Fv/Fm	1.275	Total iron content-leaf	1.280	Citrate-leaf	1.510
PbCS2	1.174	*PbCS2*	1.258	Soil pH	1.122	*PbCS2*	1.500
Total iron-leaf	1.072	Available iron-soil	1.221	Fv/Fm	1.111	Soil pH	1.325
Root activity	1.064	Root activity	1.178	SPAD	1.104	Active iron/Total iron	1.086
Soil pH	1.034	Active iron-root	1.056	Citrate-leaf	1.029		
		Citrate-leaf	1.007				

Further comparison of the PCA loadings and VIP values revealed some degree of difference in variable selection. This is primarily because PCA identifies the directions of maximum variance in the dataset, while PLS-DA VIP analysis selects the most discriminative variables for classification. Therefore, in PCA, parameters such as SPAD, citrate content, and soil active iron content may exhibit higher loadings, whereas in VIP analysis, *PbFRO2* contributes more to classification. This suggests that PCA primarily captures the overall physiological changes under iron deficiency, while VIP analysis highlights the molecular regulators that are crucial for adaptation. Overall, *PbFRO2* plays a key role at all time points, and different compensatory mechanisms may be involved in maintaining iron homeostasis during specific growth stages.

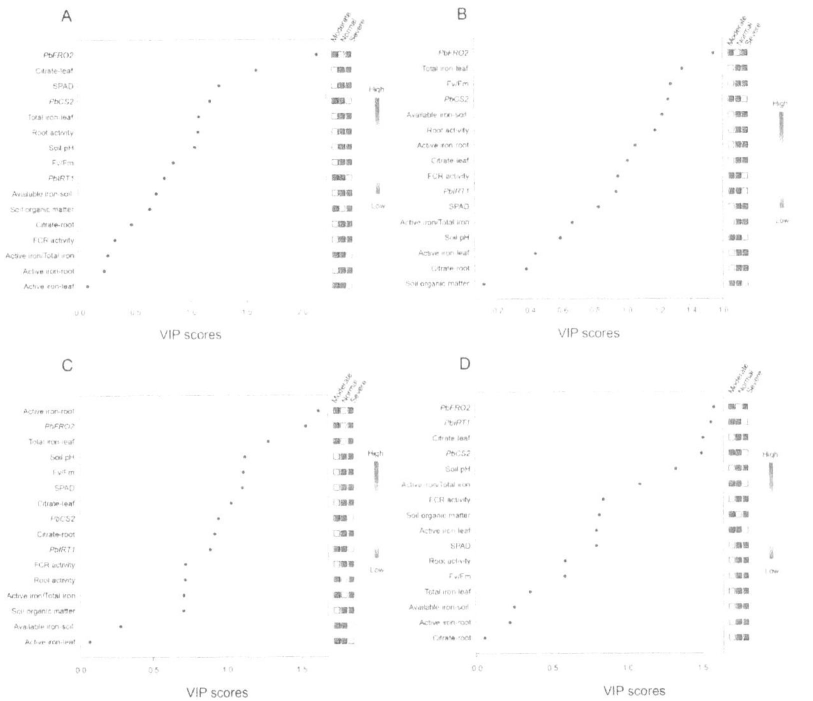

Figure 8. Partial least squares discriminant analysis (PLS-DA) with variable importance in projection (VIP) analysis of 'Yali' pears at different sampling times under different degrees of iron deficiency. (**A**) All samples grouped by chlorosis severity, regardless of sampling time. (**B**) Data from May. (**C**) Data from June. (**D**) Data from July.

3.7.3. Correlation Analysis Between Physiological Traits and Gene Expression

To further clarify the mechanism of *PbFRO2* response to iron deficiency, we conducted a correlation analysis on the dataset. Although most of the *p*-values in these analyses did not reach conventional significance levels ($p < 0.1$), the effect sizes (r values) and observed trends revealed several biologically meaningful relationships. Therefore, our study primarily focuses on descriptive analysis and trend presentation, emphasizing the magnitude and direction of effects between variables, and further validation will be needed through larger sample sizes or functional experiments.

In the analysis considering only chlorosis severity (without differentiating sampling months, see Figure 9A), the active iron content in leaves was negatively correlated with *PbFRO2* ($r = -0.127$, $p = 0.527$), *PbCS2* ($r = -0.503$, $p < 0.05$), and *PbIRT1* ($r = -0.584$, $p < 0.05$), suggesting that the reduction of active iron in leaves may induce the upregulation of these three genes. Meanwhile, the ferric reductase activity (FCR activity) showed a positive correlation with *PbFRO2* ($r = 0.018$, $p = 0.930$), *PbCS2* ($r = 0.207$, $p < 0.05$), and *PbIRT1* ($r = 0.579$, $p < 0.05$), indicating that the increased expression of these genes might enhance root ferric reductase activity, thereby promoting iron absorption.

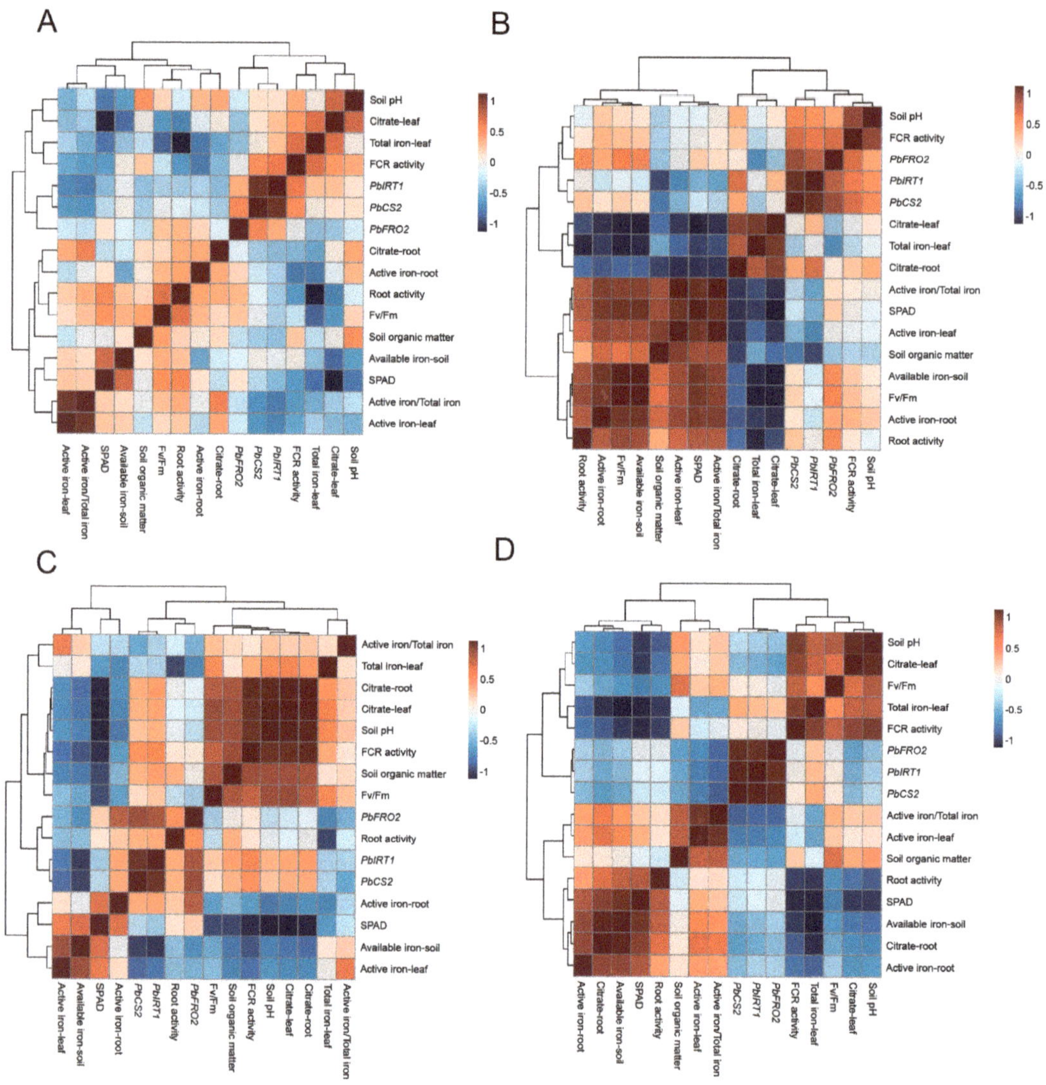

Figure 9. Correlation analysis of 'Yali' pears at different sampling times under different degrees of iron deficiency. (**A**) Grouped by chlorosis severity regardless of sampling time. (**B**) Data from May. (**C**) Data from June. (**D**) Data from July.

Subsequently, we analyzed the correlations in the dataset from May to July (Figure 9B–D), with a focus on the relationships between *PbFRO2* and leaf active iron content, root active iron content, and root FCR enzyme activity (Table 4). The results revealed that temporal changes in the relationship between *PbFRO2* and iron homeostasis suggested the existence of a stage-specific regulatory mechanism.

In the early growth stage (May) (Table 5), the strong correlation between *PbFRO2* expression and FCR activity (r = 0.619), along with its synergistic relationship with root iron accumulation (r = 0.373), aligns with the "Strategy I" iron acquisition mechanism, supporting the plant's nutritional growth requirements. However, from May (r = −0.003) to July (r = −0.582), the gradually increasing negative correlation with leaf active iron suggests that the function of this gene extends beyond simple iron absorption promotion. By June, when root iron retention peaked (r = 0.739), the newly emerging negative correlation with leaf iron (r = −0.427) and the reduction in ferric reductase activity (r = 0.071) occurred synchronously, indicating that prioritizing iron storage in the roots may be an adaptive strategy to cope with iron deficiency stress. In July, the reversal of the correlation between *PbFRO2* expression and root iron (r = −0.261), along with the exacerbation of leaf iron inhibition (r = −0.582), collectively suggested the redistribution of iron to reproductive organs.

Table 5. Effect sizes (Pearson's *r*) of *PbFRO2* expression across indicators.

Indicator	May		June		July	
	Effect Sizes	*p*-Values	Effect Sizes	*p*-Values	Effect Sizes	*p*-Values
Leaf active iron	−0.003	0.994	−0.427	0.251	−0.582	0.100
Root active iron	0.373	0.323	0.739	0.023	−0.261	0.498
FCR	0.619	0.075	0.071	0.855	−0.043	0.913

In conclusion, these seasonal dynamic characteristics indicate that *PbFRO2* coordinates iron distribution through a three-phase program: spring root iron storage dependent on reductase activity, summer root iron buffering in response to environmental stress, and reproductive phase iron redistribution accompanied by leaf iron deactivation. This gene may serve as a regulatory hub, dynamically balancing iron distribution strategies across different developmental stages. It is important to note that, although the observed large effect sizes suggest potential biological associations, the lack of statistical significance ($p > 0.1$) in this section prevents definitive conclusions. These findings should be considered exploratorily, and the research team plans to further validate them through larger sample sizes or functional experiments.

Furthermore, across various comparisons, we observed significant correlations among *PbFRO2*, *PbIRT1*, and *PbCS2* ($p < 0.05$), suggesting a synergistic interaction among these genes in regulating iron uptake and utilization. Specifically, as a key gene in ferric reduction, *PbFRO2* may enhance the bioavailability of iron by increasing the efficiency of Fe^{3+} reduction in the roots; *PbIRT1* is responsible for transporting the reduced Fe^{2+} into the plant, thereby ensuring efficient distribution of iron; meanwhile, *PbCS2* regulates the synthesis of citrate, which aids in the solubilization and translocation of iron. These significant correlations support the existence of an integrated regulatory network, indicating that pear trees coordinate multiple molecular mechanisms to optimize iron absorption and redistribution, thereby maintaining normal growth and development under iron deficiency stress.

4. Discussion and Conclusions

In this study, the active iron content in leaves of moderately and severely chlorotic groups was significantly reduced compared to the normal group, paralleling the decline in chlorophyll levels, which could be attributed to the critical role of iron in chlorophyll biosynthesis [30,31]. Consistent with the "chlorosis paradox" observed in previous studies [32], total iron content in normal, mildly, and severely chlorotic leaves exhibited irregular fluctuations, indicating that total iron cannot serve as a reliable diagnostic indicator for the iron nutritional status of 'Yali' pears.

The molecular regulatory mechanism of iron uptake in 'Yali' pears exhibits a stage-specific response to chlorosis severity. Under moderate iron deficiency, root FCR activity and *PbFRO2* expression are upregulated, consistent with the classical Strategy I mechanism of enhancing Fe^{3+} reduction and solubilization to counteract iron deficiency [13]. However, severe iron deficiency suppresses FCR activity, likely due to irreversible enzyme denaturation or cellular redox imbalance, further exacerbating iron limitation. Concurrently, the upregulation of *PbIRT1* in moderately chlorotic roots mirrors the activation of Fe^{2+} transporters observed in *Arabidopsis* [33], indicating evolutionary conservation of iron uptake mechanisms in dicots. The CS2 gene plays a pivotal role in citrate biosynthesis, exerting dual functions in iron activation and transport through citrate metabolism [34]. In this study, leaf citrate content progressively accumulated with increasing chlorosis severity, while root citrate content (excluding July samples) exhibited a similar trend.

Multivariate analyses revealed a biphasic adaptation strategy. PCA demonstrated progressive physiological divergence, with severely chlorotic leaves forming a distinct cluster separate from normal and moderately chlorotic leaves (Figure 7A), indicating threshold-dependent collapse of metabolic and molecular adjustment mechanisms. PLSDA-VIP identified *PbFRO2* as a core regulatory factor (VIP > 1.5, Table 4), whose expression strongly correlated with FCR activity ($r = 0.73$, $p < 0.01$) and active iron redistribution. Notably, transcript levels of *PbFRO2*, *PbIRT1*, and *PbCS2* peaked in moderately chlorotic plants but markedly declined under severe stress (Figure 6). This pattern suggests two potential mechanisms: (1) Plants may maintain a minimal transcriptional threshold of these genes to achieve adaptive iron homeostasis regulation; beyond this threshold, compensatory mechanisms fail, leading to irreversible physiological damage. (2) Transcriptional suppression of *PbFRO2*, *PbIRT1*, and *PbCS2* could exacerbate iron deficiency chlorosis in pear trees, creating a self-reinforcing vicious cycle.

This study deciphered the molecular regulatory mechanisms underlying iron deficiency chlorosis in 'Yali' pears through integrative analyses. Under moderate iron deficiency, the synergistic upregulation of *PbFRO2*, *PbIRT1*, and *PbCS2* likely triggered a compensatory cascade, alleviating chlorosis via enhanced Fe^{3+} reduction, Fe^{2+} transport, and citrate-mediated long-distance iron translocation. Conversely, severe iron deficiency potentially overwhelms the regulatory network capacity, inducing gene expression threshold effects, inactivation of FCR, and photosynthetic dysfunction, ultimately disrupting iron homeostasis (these mechanisms require further validation). Organ-specific citrate dynamics revealed divergent adaptive strategies: leaf citrate prioritized iron reactivation to sustain photosynthetic function, while root citrate modulated seasonal iron allocation through concentration fluctuations. These findings establish a "gene-metabolite-phenotype" tripartite framework elucidating the molecular basis of iron stress in alkaline soils, with *PbFRO2* serving as a central hub gene driving adaptive responses and offering potential targets for iron biofortification.

This work systematically characterizes the dynamic physiological and molecular adaptations during iron-deficient chlorosis in 'Yali' pear, providing novel insights into iron homeostasis regulation in perennial fruit trees and guiding iron-efficient cultivar breeding and precision fertilization in calcareous soils. However, the etiology of pear chlorosis remains dual-natured: whether it primarily stems from insufficient soil iron availability or compromised plant iron response capacity warrants resolution. To address this, follow-up studies will employ controlled iron supplementation combined with *PbFRO2* overexpression and knockout approaches to dissect the causation of chlorosis and delineate the functional network of this pivotal gene.

Author Contributions: Conceptualization, H.S. and Y.Z.; methodology, S.L.; software, M.Z. and C.W.; validation, S.L., H.W. and Y.X.; formal analysis, M.Z. and C.W.; investigation, S.L., H.W. and Y.X.; data curation, S.L.; writing—original draft preparation, S.L.; writing—review and editing, S.L., M.Z., J.Z., Y.Z. and H.S.; visualization, M.Z.; supervision, H.S. and Y.Z.; funding acquisition, H.S. All authors have read and agreed to the published version of the manuscript.

Funding: This work was supported by the National Natural Science Foundation of China (32272654), the Hebei Natural Science Foundation, China (C2023204016), the Hebei Province Introduced

Overseas-Scholar Fund, China (C20220361), and the Hebei Province Outstanding Youth Fund, China (2016, 2019).

Data Availability Statement: Data are contained within the article.

Conflicts of Interest: The authors declare no conflicts of interest.

References

1. Merry, R.; Dobbels, A.A.; Sadok, W.; Naeve, S.; Stupar, R.M.; Lorenz, A.J. Iron Deficiency in Soybean. *Crop Sci.* **2022**, *62*, 36–52. [CrossRef]
2. Zebec, V.; Lisjak, M.; Jović, J.; Kujundžić, T.; Rastija, D.; Lončarić, Z. Vineyard Fertilization Management for Iron Deficiency and Chlorosis Prevention on Carbonate Soil. *Horticulturae* **2021**, *7*, 285. [CrossRef]
3. Zang, Y.; Huang, Y.; Chang, X.; Chen, J.; Jiang, T.; Wu, Z.; Lu, L.; Tian, S. High Soil pH and Plastic-Shed Lead to Iron Deficiency and Chlorosis of Citrus in Coastal Saline–Alkali Lands: A Field Study in Xiangshan County. *Horticulturae* **2023**, *9*, 437. [CrossRef]
4. Kudirka, G.; Viršilė, A.; Sutulienė, R.; Laužikė, K.; Samuolienė, G. Precise Management of Hydroponic Nutrient Solution pH: The Effects of Minor pH Changes and Mes Buffer Molarity on Lettuce Physiological Properties. *Horticulturae* **2023**, *9*, 837. [CrossRef]
5. Guo, G.; Xiao, J.; Jeong, B.R. Iron Source and Medium pH Affect Nutrient Uptake and Pigment Content in Petunia Hybrida 'Madness Red' Cultured in Vitro. *Int. J. Mol. Sci.* **2022**, *23*, 8943. [CrossRef] [PubMed]
6. Zhang, Y.; Liu, W.-J.; Cheng, X.-H.; Du, G.-Q.; Zhang, Y.-X. Effects of Pollinizer Cultivars on the Morphogenesis and Development of Yatu and the Endogenous Hormone Content in the Base of Young Fruit Of'yali'pear (Pyrus Bretschneideri Rehd.). *J. Fruit Sci.* **2020**, *37*, 1168–1174. [CrossRef]
7. Ren, Z.-Y.; Cheng, Y.-D.; Guan, Y.-Q.; Yang, X.-Z.; Guan, J.-F. Detection of Freshness and Damage Degree of Pyrus Bretschneideri Rehd. Based on Electronic Nose. *J. Food Saf. Qual.* **2022**, *13*, 2770–2777.
8. Li, J.; Cao, X.; Jia, X.; Liu, L.; Cao, H.; Qin, W.; Li, M. Iron Deficiency Leads to Chlorosis through Impacting Chlorophyll Synthesis and Nitrogen Metabolism in *Areca catechu* L. *Front. Plant Sci.* **2021**, *12*, 710093. [CrossRef] [PubMed]
9. Rahimi, S.; Talebi, M.; Baninasab, B.; Gholami, M.; Zarei, M.; Shariatmadari, H. The Role of Plant Growth-Promoting Rhizobacteria (Pgpr) in Improving Iron Acquisition by Altering Physiological and Molecular Responses in Quince Seedlings. *Plant Physiol. Biochem.* **2020**, *155*, 406–415. [CrossRef]
10. Martín-Barranco, A.; Spielmann, J.; Dubeaux, G.; Vert, G.; Zelazny, E. Dynamic Control of the High-Affinity Iron Uptake Complex in Root Epidermal Cells1. *Plant Physiol.* **2020**, *184*, 1236–1250. [CrossRef]
11. Dubeaux, G.; Neveu, J.; Zelazny, E.; Vert, G. Metal Sensing by the Irt1 Transporter-Receptor Orchestrates Its Own Degradation and Plant Metal Nutrition. *Mol. Cell* **2018**, *69*, 953–964.e5. [CrossRef] [PubMed]
12. Han, D.-G.; Shi, Y.; Wang, B.; Liu, W.; Yu, Z.-Y.; Lv, B.-Y.; Yang, G.-H. Isolation and Preliminary Functional Analysis of Mxcs2: A Gene Encoding a Citrate Synthase in Malus Xiaojinensis. *Plant Mol. Biol. Report.* **2014**, *33*, 133–142. [CrossRef]
13. Wang, Y.; Kang, Y.; Zhong, M.; Zhang, L.; Chai, X.; Jiang, X.; Yang, X.J.A. Effects of Iron Deficiency Stress on Plant Growth and Quality in Flowering Chinese Cabbage and Its Adaptive Response. *Agronomy* **2022**, *12*, 875. [CrossRef]
14. Lucena, J.J.; Hernandez-Apaolaza, L. Iron Nutrition in Plants: An Overview. *Plant Soil* **2017**, *418*, 1–4. [CrossRef]
15. Thomas, G.W. Soil pH and soil acidity. In *Methods of Soil Analysis: Part 3 Chemical Methods 5*; John Wiley & Sons: Hoboken, NJ, USA, 1996; pp. 475–490.
16. Ma, Y.; Li, J.; Tian, M.; Liu, Y.; Wei, A. Authentication of Chinese Prickly Ash by Its2 Sequencing and the Influence of Environmental Factors on Pericarp Quality Traits. *Ind. Crops Prod.* **2020**, *155*, 112770. [CrossRef]
17. Shidan, B. *Soil Agrochemical Analysis*, 3rd ed.; Agriculture Press: Beijing, China, 2000.
18. Lv, X.; Kong, H.; Luo, Y.; Dong, D.; Liu, W.; Wu, D.; Ye, Z.; Ma, J.; Liu, D. The Impact of Magnesium on the Growth, Physiology and Quality of Tea (*Camellia sinensis* L.) Plants under Acid Stress. *Agronomy* **2024**, *14*, 767. [CrossRef]
19. Zhou, J.; Yuan, W.; Di, B.; Zhang, G.; Zhu, J.; Zhou, P.; Ding, T.; Qian, J. Relationship among Electrical Signals, Chlorophyll Fluorescence, and Root Vitality of Strawberry Seedlings under Drought Stress. *Agronomy* **2022**, *12*, 1428. [CrossRef]
20. Zheng, S.J.; Tang, C.; Arakawa, Y.; Masaoka, Y. The Responses of Red Clover (*Trifolium pratense* L.) to Iron Deficiency: A Root Fe(Iii) Chelate Reductase. *Plant Sci.* **2003**, *164*, 679–687. [CrossRef]
21. Lin, L.; Li, Z.; Wu, C.; Xu, Y.; Wang, J.; Lv, X.; Xia, H.; Liang, D.; Huang, Z.; Tang, Y. Melatonin Promotes Iron Reactivation and Reutilization in Peach Plants under Iron Deficiency. *Int. J. Mol. Sci.* **2023**, *24*, 16133. [CrossRef]
22. Lucarini, M.; Di Lullo, G.; Cappelloni, M.; Lombardi-Boccia, G. In Vitro Estimation of Iron and Zinc Dialysability from Vegetables and Composite Dishes Commonly Consumed in Italy: Effect of Red Wine. *Food Chem.* **2000**, *70*, 39–44. [CrossRef]
23. Zhang, C.; Wang, Y.; Li, Y. Effect of Iron Deficiency on Contents of Nutrient Elements and Photosynthetic Characteristic of Leaves in Pear Trees. *J. Northwest For. Univ.* **2002**, *17*, 9–11.
24. Zhang, R.; Yang, P.; Liu, S.; Wang, C.; Liu, J. Evaluation of the Methods for Estimating Leaf Chlorophyll Content with Spad Chlorophyll Meters. *Remote Sens.* **2022**, *14*, 5144. [CrossRef]

25. Zhao, J. Photosynthetic Characteristics Research and Fruit Quality Determination of Different Planting Density on Korla Fragrant Pear. Master's Thesis, Xinjiang Agricultural University, Urumqi, China, 2014. (In Chinese).
26. Nsiri, K.; Krouma, A. The Key Physiological and Biochemical Traits Underlying Common Bean (*Phaseolus vulgaris* L.) Response to Iron Deficiency, and Related Interrelationships. *Agronomy* **2023**, *13*, 2148. [CrossRef]
27. Rewald, B.; Meinen, C. Plant Roots and Spectroscopic Methods—Analyzing Species, Biomass and Vitality. *Front. Plant Sci.* **2013**, *4*, 393. [CrossRef]
28. Li, X. The Response Mechanism of Pyrus Betulifolia Root System to Iron Deficiency Stress and the Expression Analysis of Related Resistance Genes. Ph.D. Thesis, Hebei Agricultural University, Baoding, China, 2016. (In Chinese).
29. Tahjib-Ul-Arif, M.; Zahan, M.I.; Karim, M.M.; Imran, S.; Hunter, C.T.; Islam, M.S.; Mia, M.A.; Hannan, M.A.; Rhaman, M.S.; Hossain, M.A.; et al. Citric Acid-Mediated Abiotic Stress Tolerance in Plants. *Int. J. Mol. Sci.* **2021**, *22*, 7235. [CrossRef]
30. Kong, W.-L.; Wu, X.-Q.; Zhao, Y.-J. Effects of Rahnella Aquatilis Jz-Gx1 on Treat Chlorosis Induced by Iron Deficiency in Cinnamomum Camphora. *J. Plant Growth Regul.* **2019**, *39*, 877–887. [CrossRef]
31. Bityutskii, N.P.; Yakkonen, K.L.; Lukina, K.A.; Semenov, K.N. Fullerenol Increases Effectiveness of Foliar Iron Fertilization in Iron-Deficient Cucumber. *PLoS ONE* **2020**, *15*, e0232765. [CrossRef] [PubMed]
32. Gao, Z.; Zhao, C.; Dong, D.; Liu, S.; Wen, X.; Gu, Y.; Jiao, L. Visualizing Changes of Metabolites During Iron Deficiency Chlorosis in Field-Grown Pear Leaves Using Micro-Raman Spectral Imaging. *Front. Plant Sci.* **2022**, *13*, 1079660. [CrossRef]
33. Gong, Q.; Zhou, M.; Li, X.; Guo, Y. Transcription Factor Myb8 Regulates Iron Deficiency Stress Response in Arabidopsis. *Plant Sci.* **2024**, *340*, 111973. [CrossRef]
34. Zhang, X.; Wei, X.; Ali, M.M.; Rizwan, H.M.; Li, B.; Li, H.; Jia, K.; Yang, X.; Ma, S.; Li, S.; et al. Changes in the Content of Organic Acids and Expression Analysis of Citric Acid Accumulation-Related Genes During Fruit Development of Yellow (*Passiflora edulis* f. *flavicarpa*) and Purple (*Passiflora edulis* f. *edulis*) Passion Fruits. *Int. J. Mol. Sci.* **2021**, *22*, 5765. [CrossRef]

MDPI AG
Grosspeteranlage 5
4052 Basel
Switzerland
Tel.: +41 61 683 77 34

Horticulturae Editorial Office
E-mail: horticulturae@mdpi.com
www.mdpi.com/journal/horticulturae

www.ingramcontent.com/pod-product-compliance
Lightning Source LLC
Chambersburg PA
CBHW062037220926
51850CB00010B/2581